292
Current Topics in Microbiology and Immunology

Z.F. Fu (Ed.)

The World of Rhabdoviruses

With 27 Figures and 7 Tables

Springer

Professor Dr. Zhen F. Fu, DVM, PhD
Department of Pathology
College of Veterinary Medicine
University of Georgia
Athens, GA 30602
USA
e-mail: zhenfu@vet.uga.edu

Cover figure by Zhen F. Fu

Library of Congress Catalog Number 72-152360

ISBN-10 3-540-24011-X Springer Berlin Heidelberg New York
ISBN-13 978-3-540-24011-2 Springer Berlin Heidelberg New York

Springer is a part of Springer Science+Business Media
springeronline.com

Printed in The Netherlands

Editor: Dr. Paul Roos, Heidelberg
Desk editor: Anne Clauss, Heidelberg
Production editor: Nadja Kroke, Leipzig
Cover design: design & production GmbH, Heidelberg
Typesetting: LE-TEX Jelonek, Schmidt & Vöckler GbR, Leipzig
Printed on acid-free paper SPIN 11001713 21/3150/YL – 5 4 3 2 1 0

Preface

Rhabdoviruses have a very wide host range and have been isolated from plants, insects, and almost from all vertebrates including fish and primates. The Rhabdoviridae family consists of six genera which have all been associated with diseases, either in animals or plants. While rhabdoviruses that are the etiological agents of human diseases can cause serious public health problems, other members of this family that infect domestic livestock and agricultural plants can also cause enormous economic loss. Despite the significance of rhabdoviruses for public health and agriculture, the last book exclusively devoted to these viruses was published in 1987. The first classification of rhabdoviruses was based on the distinct bullet-shaped morphology that is characteristic of members of this family. However, using modern gene technology, which allows the complete analysis of entire viral genomes, it was recently found that different rhabdoviruses exhibit not only morphological similarities but are also genetically related. In this respect, the development of reverse genetics technology and its use for studying the regulation of viral transcription and replication, deciphering the pathogenic mechanisms, and developing vaccines and gene therapy vectors probably resulted in the most significant progress in rhabdovirus research during the past 15 years.

This volume is intended to review the unique and common features of rhabdoviruses, particularly their morphological, molecular, and pathogenic characteristics, and their phylogenetic relationships.

The chapter by Fu reviews the common characteristics of rhabdoviruses, particularly from the viewpoint of phylogenetic relationships. The chapter by Dietzschold et al. summarizes the latest findings on the molecular pathogenic mechanisms of rabies. Hoffmann et al. discuss the molecular epidemiology and evolution of fish rhabdoviruses. Walker describes the epidemiology of and the diseases caused by ephemeroviruses. Warrilow summarizes how a new genus of rhabdoviruses, the Australian bat lyssaviruses, was discovered and how this genus of viruses is maintained and transmitted by insect- as well as fruit-eating bats. Redinbaugh and Hogenhout summarize recent findings on plant rhabdoviruses.

The development of reverse genetics in the 1990s has enabled the manipulation of the rhabdoviral genome and thus revolutionized the research field for rhabdoviruses. The contribution by Finke and Conzelmann reviews the development and establishment of this technology for vesicular stomatitis virus and rabies virus. Furthermore, this chapter also updates the progress of using this technology for the development of vaccines and gene therapy vectors. In addition, Brémont describes the establishment of a reverse genetics system for fish rhabdoviruses. This technology is also a powerful tool for the investigation of pathogenic mechanisms by which rhabdoviruses induce diseases in respective hosts. An example is the chapter by Dietzschold et al. in which different rabies viruses were constructed to decipher the contribution of each of the proteins in the induction of rabies. With either the minigenome or the infectious clones in the vesicular stomatitis virus system, many of the *cis-* and *trans-*elements important in the process of transcription and replication have been identified and/or confirmed, some of which are described in the chapter by Fu.

I would like to take this opportunity to thank Dr. Hilary Koprowski for constant encouragement and help during the editing of this volume and all of the contributors to this volume for their patience and enthusiasm.

March 2005 *Zhen F. Fu*

List of Contents

List of Contributors

(Their addresses can be found at the beginning of their respective chapters.)

CTMI (2005) 292:1–24

Genetic Comparison of the Rhabdoviruses from Animals and Plants

Z. F. Fu

Department of Pathology, University of Georgia, 501 D. W. Brooks Drive, Athens GA, 30606, USA
zhenfu@vet.uga.edu

Abstract There are more than 160 viral species in the *Rhabdovidae* family, most of which can be grouped into one of the six genera including *Vesiculovirus*, *Lyssavirus*, *Ephemerovirus*, *Novirhabdovirus*, *Cytorhabdovirus*, and *Nucleorhabdovirus*. These viruses are not only morphologically similar but also genetically related. Analysis of viral genes shows that rhabdoviruses are more closely related to each other than to viruses in other families. With the development of reverse genetics, the functions of many *cis*- and *trans*-elements important in the process of viral transcription and replication have been clearly defined such as the leader, trailer, and the intergenic sequences. Furthermore, it has been shown that there are two entry sites for the RNA-dependent RNA polymerase: 3′ entry for leader synthesis and RNA replication, and direct entry at the N gene start sequence for transcription of the monocistronic mRNAs.

1 Introduction

Rhabdoviruses are a large family of viruses that can infect a wide range of hosts (Walker et al. 2000). Rhabdovirus has been isolated from plants, invertebrate and vertebrate animals (for reviews, see Wagner 1987; Dietzschold et al. 1996; Rose and Whitt 2000). To date there are more than 160 species of rhabdoviruses isolated (Walker et al. 2000), with new viruses being discovered constantly (Amelia et al. 2002; Kuzmin et al. 2003; Mork et al. 2004). Some of the rhabdoviruses induce severe diseases in humans and animals, including fish (Dietzschold et al. 1996; Brown 1987; Ahne et al. 2002). Others cause plant diseases, particularly in agricultural crops (Jackson et al. 1987). Previously viruses were classified as rhabdoviruses because of their distinct bullet-shaped morphology (Wagner 1987). With recent development of genetic analysis, these viruses are clearly not only morphologically similar, but also genetically related (Walker et al. 2000; Warrilow et al. 2002). This paper is intended to present the genetic relatedness of all the rhabdoviruses regardless the sources of isolation. Furthermore, recent advances in reverse genetics have defined more clearly the *cis*- and *trans*-elements that are important regulators of rhabdoviral transcription and replication.

2 Classification of Rhabdoviruses

The family of Rhabdoviridae has been classified in the order of Mononegavirales together with *Paramyxoviridae*, *Filoviridae*, and *Bornaviridae* (Murphy 1996). To date, more than 160 species of rhabdoviruses have been reported, most of which can be classified or tentatively classified in six genera and others have yet to be assigned (Walker et al. 2000). The six genera include *Vesiculovirus*, *Lyssavirus*, *Ephemerovirus*, *Novirhabdovirus*, *Cytorhabdovirus*, and *Nucleorhabdovirus*. Many of the well-known and characterized vesiculoviruses include VSV Indiana virus (VSIV), and VSV New Jersey virus (VSNJ). The less characterized vesiculoviruses include Alagoas virus (VSAV), Carajas virus (CJSV), Chandipura virus (CHPV), Cocal virus (COCV), Isfahan virus (ISFV), Maraba virus (MARAV), and Piry virus (PIRYV) (Walker et al. 2000). The Lyssavirus genus includes rabies and rabies-related viruses. Molecular comparison has divided lyssaviruses into seven genotypes (Bourhy et al. 1993; Gould et al. 1998). All the classical rabies viruses (RABV) belong to genotype 1; Lagos bat virus (LBV) genotype 2; Mokola virus (MOKV) genotype 3; Duvenhage virus (DUVV) genotype 4, European bat Lyssavirus

1 (EBLV-1) genotype 5, European bat Lyssavirus 2 (EBLV-2) genotype 6; and the newly discovered Australian bat lyssavirus (ABLV) belong to genotype 7. The Ephemerovirus genus includes bovine ephemeral fever virus (BEFV), Adelaide River virus (ARV), and Berrimah virus (BRMV) (Walker et al. 2000). The genus of Novirhabdovirus also has three well-defined species that infect aquatic hosts, i.e., infectious hematopoietic necrosis virus (IHNV), viral hemorrhagic septicemia virus (VHSV), and hirame rhabdovirus (HIRRV) (Kurath et al. 1997; Walker et al. 2000). The last two genera in the family of Rhabdoviridae are the viruses infecting plants, the Cytorhabdovirus and the Nucleorhabdovirus (Walker et al. 2000). Well-characterized cytorhabdoviruses include lettuce necrotic yellow virus (LNYV), barley yellow striate mosaic virus (BYSMV), broccoli necrotic yellow virus (BNYV), and Northern cereal mosaic virus (NCMV) (Jackson 1987; Wetzel et al. 1994). Well-characterized nucleorhabdoviruses include eggplant mottled dwarf virus (EMDV), maize mosaic virus (MMV), rice yellow stunt virus (RYSV), sonchus yellow net virus (SYNV), and potato yellow dwarf virus (PYDV) (Jackson et al. 1987; Martins et al. 1998).

3
Diseases Caused by Rhabdoviruses

The importance of rhabdoviruses is that these viruses cause severe diseases in plants and animals, including humans. For example, rabies virus causes rabies in humans as well as in all warm-blooded animals, with almost 100% mortality (Dietzschold et al. 1996). VSV causes a disease that is clinically similar to that of foot-and-mouth disease in cattle and pigs (Brown 1987). Bovine ephemeral fever virus causes a disabling viral disease of cattle and water buffalo (St George 1990). Spring viremia of carp virus (SVCV) causes a severe hemorrhagic disease of cyprinids (Ahne et al. 2002). Plant rhabdoviruses cause many plant diseases including, but not limited to, maize mosaic, rice transitory yellowing, potato yellow dwarf, and lettuce necrotic yellows (Jackson et al. 1987). Thus, rhabdoviruses not only cause diseases in humans, thus presenting public health problems, but also induce diseases in domestic and wildlife animals and fish, as well as in agricultural plants, causing economic losses.

4 Transmission of Rhabdoviruses

Many of the rhabdoviruses are transmitted by insect vectors. Plant-infecting rhabdoviruses are transmitted to their plant hosts by insect vectors, including aphids, planthoppers and leafhoppers (Jackson 1987). Several animal rhabdoviruses, for example, ephemeroviruses, and some vesiculoviruses, are also transmitted by insect vectors (Wagner 1987; Walker et al. 2000). Actually, many of the vesiculoviruses and ephemeroviruses were isolated initially from insects. For example, Cocal virus was isolated from dust mites (Jonkers et al. 1964), and Sigma virus was recognized first as a congenital infection of *Drosophila* (Printz 1973). Others can readily infect mosquitoes or mosquito cells (Gillies and Stollar 1980). Recently it has been proposed that VSV is transmitted by mosquitoes from wild to domestic animals on Ossabaw Island, although transmission by insect vectors among the wild animal population is not clear (Stallknecht 2000). The widespread ability of rhabdoviruses to infect insects has led to the suggestion that the family of Rhabdoviridae evolved from an ancestral insect virus (Nault 1997) and that the host range of rhabdoviruses is largely determined by the insect host (Hogenhout et al. 2003).

Among the rhabdoviruses, only Lyssavirus and fish rhabdoviruses (Novirhabdovirus and some vesiculoviruses) are not maintained by insect hosts. Previously, Obodhiang and Kotonkan viruses that were classified into the Lyssavirus genus were isolated from insects (Shope and Tesh 1987). These viruses are now listed as unassigned rhabdoviruses because serological and molecular data link them to viruses in the genus Ephemerovirus (Walker et al. 2000). Although it has been known for a long time that the rabies virus life cycle is maintained by carnivores (Fu 1997), now it is known that almost all the viruses in the Lyssavirus genus can be transmitted by bats. RABV (genotype 1) transmitted by vampire bats in South America has caused major economic loss in the cattle industry (Diaz et al. 1994). From the beginning of the 1990s, rabies virus strains normally circulating in the insectivorous bat population have been responsible for most human rabies cases in the United States (Krebs et al. 2000; Messenger et al. 2002; Morimoto et al. 1996). Lyssavirus genotypes 2 (LBV), 4 (DUVV), 5 (EBLV-1), and 6 (EBLV-2) have all been isolated from bats (Shope 1982). The newly discovered ALBV (genotype 7) was maintained by insectivorous as well as fruit-eating bats, the flying foxes (Gould et al. 2002). Recently serological evidence of bat rabies has been reported in the Philippines (Arguin et al. 2002) and novel lyssaviruses have been isolated in bats in Central Asia (Kuzmin et al. 2003). Obviously, further studies are needed to understand the role of bats in the evolution and transmission of lyssaviruses.

5
Genome Organization of Rhabdoviruses

Rhabdovirus genomes are among the simplistic viruses and most of them contain only five genes in the order 3′ N-P-M-G-L 5′ (Walker et al. 2000; Rose and Whitt 2000). The very 3′ end of the genome is the leader sequences and the very 5′ end of the untranslated region is the trailer sequences (Rose and Whitt 2000). Between each of these viral genes lies the intergenic sequence. The leader, trailer, and the intergenic sequences play important roles in the process of viral transcription and replication (see Sect. 3.1). In addition to these five structural proteins, two small proteins have been detected in virus-infected cells that are translated from a second reading frame within the P gene of the vesiculoviruses (Herman 1986; Spiropoulou and Nichol 1993) and lyssaviruses (Chenik et al. 1995). Some fish rhabdoviruses have an extra small gene between G and L (Kurath et al. 1985). Plant rhabdoviruses can have extra genes between P and M (Chen et al. 1998) and between G and L (Huang et al. 2003). Recently Tanno et al. (2000) reported four extra genes between P and M in Northern cereal mosaic virus. Ephemeroviruses encode extra genes but these genes were between G and L. So far six extra genes, including a second nonstructural glycoprotein (G_{NS}) and five smaller proteins ($\alpha 1$, $\alpha 2$, $\alpha 3$, β and γ), were found between the G and L genes in ephemeroviruses (Walker et al. 1992; Wang et al. 1994; McWilliam et al. 1997). The Sigma virus of *Drosophila* also has three extra genes between N and G as well as a 33-nucleotide overlap of the G gene with the preceding gene (Landes-Devauchelle et al. 1995; Teninges et al. 1993). However, the functions of these extra gene products are still unknown.

The rhabdovirus nucleoprotein (N) serves the critical function of encapsidating the genomic RNA into an RNase-resistant core that is the template for both transcription and replication (Banerjee and Chattopadhyay 1990; Blumberg et al. 1983; Wertz et al. 1987; Wunner 1991, Yang et al. 1998). By encapsidating the genomic RNA, N is thought to regulate the switch from transcription to replication. It has been calculated from the length of the genome and the number of N molecules per virion that each N molecule would cover about nine nucleotides of RNA (Wunner 1991), which has been demonstrated when the N is expressed alone (Schoehn et al. 2001; Green et al. 2000). This N–RNA complex interacts with the P–L polymerase complex during transcription and replication (Banerjee and Chattopadhyay 1990; Wertz et al. 1987). The phosphoprotein (P), on one hand, binds to N, to confer the specificity of N encapsidation of genomic RNA (Banerjee et al. 1989; Yang et al. 1998). On the other hand, P in combination with the L protein forms the RNA-dependent RNA polymerase (RdRp). The matrix protein (M) interacts with the nucleocapsid and the cytoplasmic portion of the G, thereby facilitating virus assembly and budding (Walker et al. 2000; Wunner 1991). The M protein also plays an important role in regulating viral RNA transcription (Finke and Conzelmann 2003). The glycoprotein (G) is the only surface protein for rhabdoviruses (Rose and Whitt 2000), and it plays important roles for binding to cellular receptors, thus initiating virus infection (Dietzschold et al. 1996; Walker et al. 2000). The large protein (L) is the major component of the RdRp, which is responsible for copying the N–RNA template to produce mRNA, or complete antigenomic and genomic RNA (Baltimore et al. 1970; Emerson and Yu 1975; Rose and Whitt 2000). In addition, the L also plays roles in mRNA capping, methylation of 5′ cap structures, and polyadenylation (Baltimore et al. 1970; Banerjee 1987; Rose and Whitt 2000).

6
Genetic Comparison of Rhabdoviruses

Rhabdoviruses are not only morphologically similar, but also share serological cross-reactions among some of them. Polyclonal antisera to BEFV and ARV showed cross-reactivity with RABV N protein (Walker et al. 1994). Low-level cross-reactions have also been reported between ephemeroviruses, lyssaviruses and several unclassified rhabdoviruses isolated from cattle or insects (Calisher et al. 1989). The cross-reactivity possibly reflects the genetic relatedness. Genetic comparison has been carried out previously, particularly between one or two genes of different viruses (Gallione and Rose 1983; Tordo et al. 1986a; Poch et al. 1989, 1990). Relatedness between different rhabdoviruses was observed. Recently full-length sequences have been obtained from many of the rhabdoviruses (Iverson and Rose 1981; Schubert et al. 1985; Tordo et al. 1988; Conzelmann et al. 1990; Schutze et al. 1995, 1999; Tanno et al. 2000; Ito et al. 2001; Hoffmann et al. 2002), which makes global genetic comparison possible. The full-length of rhabdoviral genomic RNA is about 11–15 kb and comparison of each of the five structural genes among the different rhabdoviruses is presented in the following sections.

6.1
N Gene

Since N is the first gene transcribed from the genome, it has been sequenced from most of the existing rhabdoviruses. Genetic comparison of N sequences from different rhabdoviruses has been used most often for evolution and classification studies among rhabdoviruses and for relatedness between rhabdovirus and other viruses (Tordo et al. 1986a; Barr et al. 1991; Bourhy et al. 1993; Wang et al. 1995; Amengual et al. 1997; Walker et al. 2000). The N sequences are highly homologous within each of the genera as well as among all the rhabdoviruses. Using a relatively conserved region of the N protein (119 amino acids), Walker et al. (2000) constructed a universal phylogeny of rhabdoviruses, which indicates that the N sequences from different rhabdoviruses are more closely related than the N sequences from other negative-stranded nonsegmented RNA viruses (for example, human paramyxovirus type 1). We also analyzed the N sequences from different genera of the rhabdoviruses by using the Phylip Package developed in the Department of Genome Sciences at University of Washington (http://evolution.genetics.washington.edu/phylip/doc/main.html references), which include a combination of Clustalw, SEQBOOT, PROTDIST, NEIGHBOR (Neighbor-Joining method of Saitou and Nei 1987), CONSENSE, and DRAWTREE. As shown in Fig. 1A, viruses in each of the genera are clustered together and separated from the others, similar to that described by Walker et al. (2000). Furthermore, common motifs have been found in the N sequences between rhabdoviruses and other negative-stranded nonsegmented RNA viruses, which may be involved in protein-RNA and protein–protein interactions in the virus nucleocapsid (Barr et al. 1991).

6.2 L Gene

The L genes are relatively conserved not only among the rhabdoviruses, but also among the negative-stranded nonsegmented RNA viruses. Six conserved domains (blocks I–VI) have been found (Poch et al. 1990) in which motifs related to enzymatic functions important for viral replication reside. Block I is critical for multiple polymerase function (Chandrika et al. 1995) and there is an invariant tripeptide GHP within this region. Block II is rich in basic residues and may play a role in RNA recognition or nucleotide binding (Muller et al. 1994; Smallwood et al. 1999). Block III has four conserved motifs, A–D, some of which are present in all known polymerases, implying the critical function of these motifs. For example, mutation of the GDN core sequence of motif C blocks viral transcription and replication (Schnell and Conzelmann 1995). Block IV is rich in proline residues among all the negative-stranded RNA viruses and may be involved in nucleotide binding. Block V has invariant cysteine and histidine residues that may play a catalytic role via metal binding. Block VI contains a glycine-rich motif (GXGXG) that

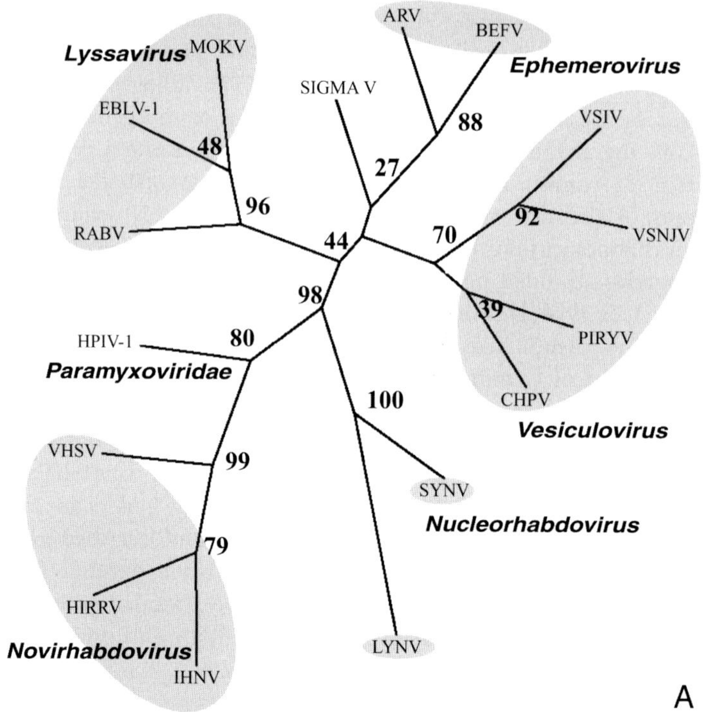

Fig. 1A–E Phylogenetic relationships between different genera of the rhabdoviruses. Rhabdoviral N (**A**), L (**B**), P (**C**), M (**D**), and G (**E**) genes were analyzed using the Phylip Package developed in the Department of Genome Sciences at the University of Washington (http://evolution.genetics.washington.edu/phylip/doc/main.html references), which include a combination of Clustalw, SEQBOOT, PROTDIST, NEIGHBOR, CONSENSE, and DRAWTREE. Wherever possible, sequences from each genus were included. Human parainfluenza virus 1 (HPIV-1) was used as the outgroup for the analyses. *ABLV*, Australia bat lyssaviruses; *ARV*, Adelaide River virus; *BEFV*, bovine ephemeral fever virus; *CHPV*, Chandipura virus; *EBLV-1*, European bat Lyssavirus 1; *HIRRV*, hirame rhabdovirus; *IHNV*, infectious hematopoietic necrosis virus; *LNYV*, lettuce necrotic yellow virus; *MOKV*, Mokola virus; *PIRYV*, Piry virus; *NCMV*, Northern cereal mosaic virus; *SIGMAV*, Sigma virus; *RABV*, rabies viruses; *SYNV*, sonchus yellow net virus; *VHSV*, viral hemorrhagic septicemia virus; *VSIV*, VSV Indiana virus; *VSNJ*, VSV New Jersey virus

may be important for polyadenylation or protein kinase activities. Using the conserved domain III sequences, Warrilow et al. (2002) analyzed the phylogenetic relationship between all the genera within the Rhabdoviridae family and found that each of the genera is clustered together. Furthermore, these authors reported that genera containing viruses that infect terrestrial

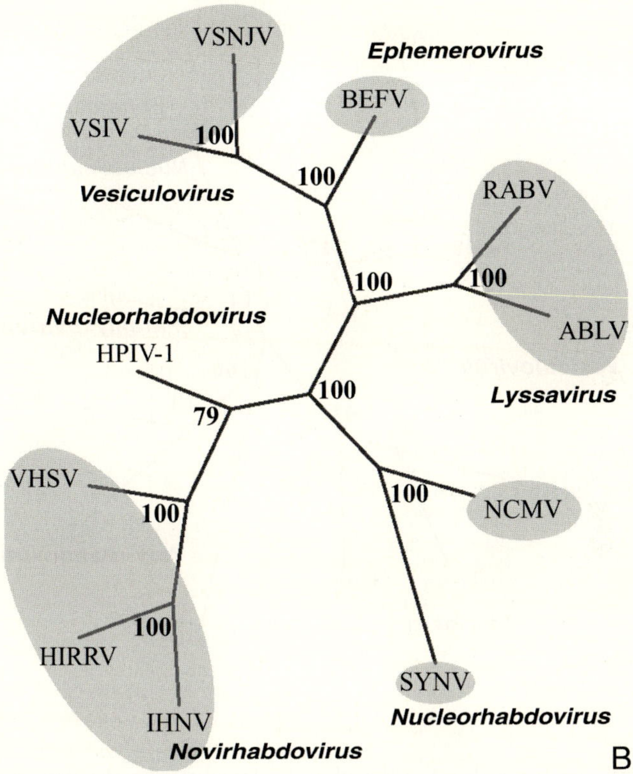

Fig. 1A–E (continued)

animals (lyssaviruses, vesiculoviruses, and ephemeroviruses) clustered more closely than those infecting fish and plants. Strong bootstrap support (>95) was found for the clustering at each of the nodes separating the genera. We analyzed the complete L sequences of representatives from each genus by using the Phylip Package and similar findings as those reported by Warrilow et al. (2002) were obtained (Fig. 1B).

6.3
P, M, and G Genes

Although sequence similarities have been reported for P, M, and G genes within some of the rhabdoviruses (Gallione and Rose 1983; Rayssiguier et al. 1986; Larson and Wunner; 1990), these sequences have not been compared among all the rhabdoviruses. Here we compared the P, M, and G sequences

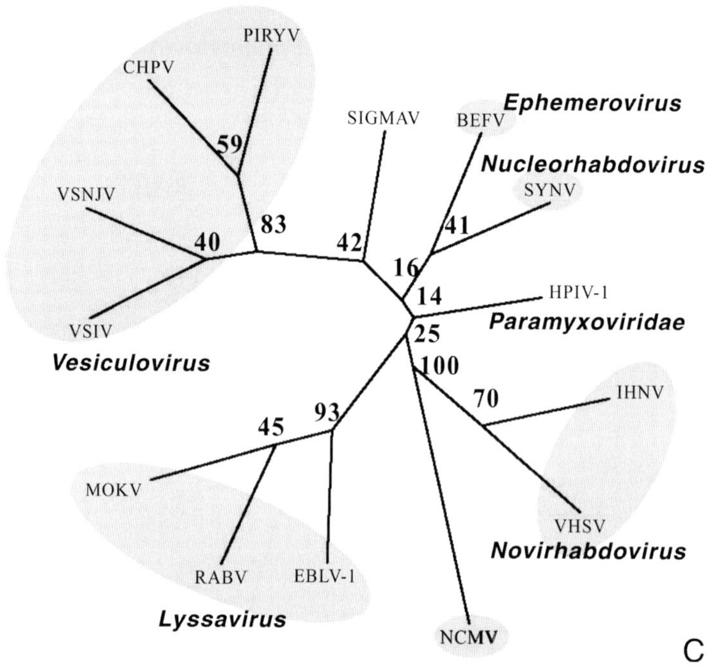

Fig. 1A–E (continued)

from all the genera within the rhabdovirus family using the Phylip Package. As shown in Fig. 1C–E, viruses in each of the genera are clustered together for each of the genes, just like the N and L genes. However, viruses that infect terrestrial animals (lyssaviruses, vesiculoviruses, and ephemeroviruses) are not always clustered together and separate from those infecting fish and plants. It is also interesting to note that the viruses transmitted by insect cells are clustered together (vesiculoviruses, ephemeroviruses, and Sigma virus) for the G genes (Fig. 1E). It is tempting to speculate that the similarities of the G may contribute to the ability of these viruses to infect insects.

7
Regulation of Rhabdoviral Transcription and Replication

The single-strand, negative-sense genomic RNA of rhabdoviruses acts as a template for both transcription and replication. During the transcription process, a positive-strand leader RNA and five or more monocistronic mRNAs are synthesized (Wertz et al. 1987, Banerjee and Chattopadhyay 1990; Wunner

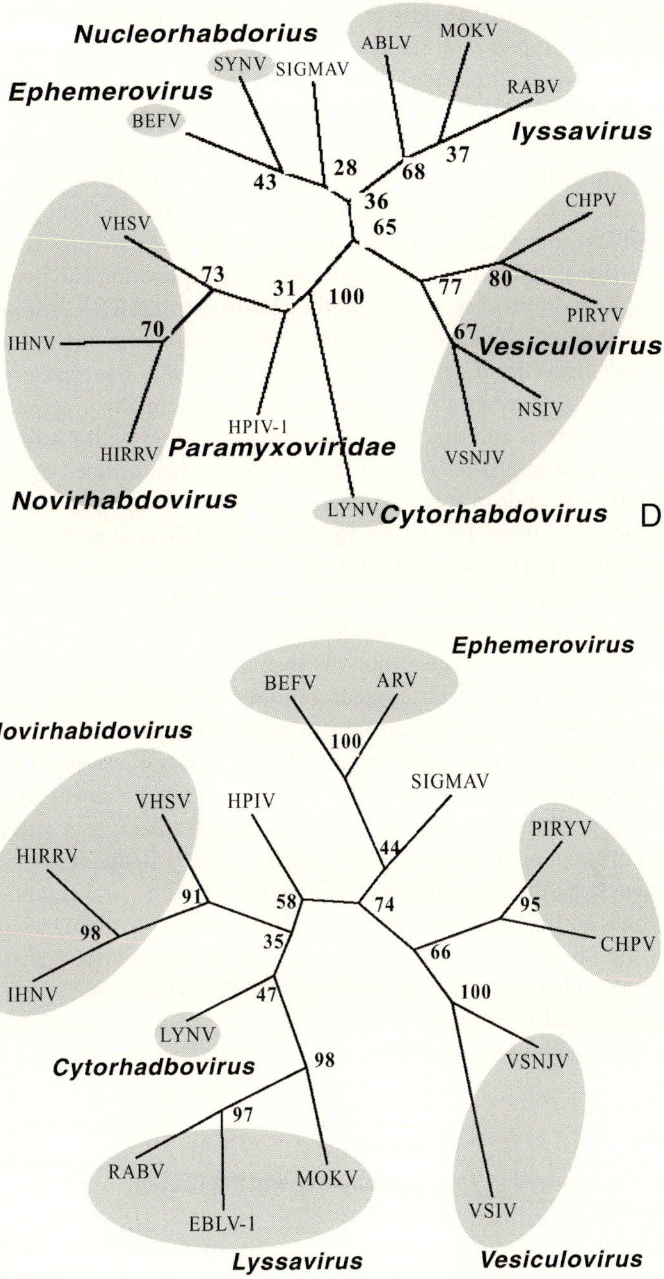

Fig. 1A–E (continued)

1991; Rose and Whitt 2000; Barr et al. 2002). Like cellular mRNA, rhabdoviral mRNAs are capped and methylated at the 5′ end and polyadenylated at the 3′ end. However, the leader RNA is neither capped nor polyadenylated. Rhabdoviral transcription initiates at the 3′ end of the genome in an obligatorily sequential manner (Abraham and Banerjee 1976; Ball and White 1976; Rose and Whitt 2000). However, the mRNAs were not produced in equimolar amounts, rather their abundance attenuated with distance from the 3′ promoter (Villarreal et al. 1976; Iverson and Rose 1981). As soon as these transcripts are translated into viral proteins, genome replication can begin. During the replication process, a full-length, positive-strand RNA (antigenome) is first synthesized, which then serves as the template for the synthesis of progeny negative-strand genomic RNA (Banerjee 1987; Banerjee and Chattopadhyay 1990; Wunner 1991). Both the genome and the antigenome are encapsidated with N and this RNA–N complex, together with P and L, forms the ribonucleoprotein (RNP) complex. The RNP is then utilized as templates for subsequent rounds of transcription, replication, or assembled into infectious particles (Wunner 1991; Rose and Whitt 2000; Barr et al. 2002).

Rhabdovirus, particularly VSV, has been used as the prototypic virus for studying the regulation of viral transcription and replication among the nonsegmented and single-stranded RNA viruses (see Barr et al. 2002 and references therein). It has been proposed that it is the availability of soluble N protein to encapsidate the nascent genomic and antigenomic RNA that switches the virus from transcription to replication (Blumberg et al. 1983; Wertz et al. 1987). In addition, many *cis*- (leader and trailer sequences, intergenic sequences) and *trans*- (viral and cellular proteins) elements have been proposed to participate in the regulation of viral transcription and replication. The most important advance in the past decade is the development of reverse genetics, which made manipulation of the rhabdoviral genome possible (Pattnaik and Wertz 1990, 1991; Schnell et al. 1994). With this technology, many of the *cis*- and *trans*-elements in the regulation of viral transcription and replication have been delineated. Recently, Barr et al. (2002) described in detail the *cis*- and *trans*-elements involved in the transcriptional control of VSV. Only the highlights are summarized in the following section.

7.1
Cis-elements Involved in Viral Transcription and Replication

The 3′ leader sequences of the genome act as a promoter for both transcription and replication (Banerjee and Chattopadhyay 1990; Wunner 1991; Rose and Whitt 2000). The 5′ trailer sequence of the genome acts exclusively as an origin of replication. However, the complementary sequence of the trailer region, i.e., the 3′ end of the antigenome, acts as the promoter for synthesis of full-length negative-sense genomes. Using reverse genetics, it is now confirmed that the genomic termini of VSV were multifunctional, containing essential signals for encapsidation of RNA by N, binding of polymerase to template, transcription, replication, and assembly of infectious particles (Pattnaik et al. 1995). The 3′ and the 5′ sequences are complementary to a certain extent in all the sequenced rhabdoviruses (Iverson and Rose 1981; Schubert et al. 1985; Tordo et al. 1988; Conzelmann et al. 1990; Schutze et al. 1995, 1999; Tanno et al. 2000; Hoffmann et al. 2002). From the studies of VSV, it has been shown that the extent of complementarity between the genomic termini affected the use of the template for transcription or replication (Wertz et al. 1994) because increasing the extent of complementarity between the termini increased replication and ultimately decreased transcription. However, Li and Pattnaik (1997) reported that deletion of nucleotides 25–45 from both termini of DI RNA but maintaining the length of terminal complementarity reduced replication by about 20-fold, which may indicate that the presence of specific sequences rather than the extent of complementarity at the termini determines the efficiency of replication.

Other *cis*-elements include the highly conserved sequences at the beginning and end of each gene and the intergenic sequence (Colonno and Banerjee 1978; Rose 1980; Tordo et al. 1986b; Walker et al. 2000). Recent findings demonstrated that these *cis*-elements are important in regulation of viral transcription (Barr et al. 2002 and references therein). The arrangement of these sequences is similar in the family of rhabdoviruses and these sequences are conserved among viruses within each genus (Wunner 1991; Rose and Whitt 2000; Walker et al. 2000). For example, each of the internal gene junctions in VSV comprised the sequence 3′...AUACUUUUUUU *G/CA*UUGUCNNAG ... 5′ (Rose and Whitt 2000; Barr et al. 2002). In addition, there is a leader-N gene junction sequence that is present neither in the leader RNA nor the N mRNA (Wunner 1990; Rose and Whitt 2000; Walker et al. 2000). Deletion of these nucleotides in VSV resulted in the abrogation of mRNA synthesis from a subgenomic replicon (Whelan and Wertz 1999), although point mutations engineered throughout this region had little effect on mRNA synthesis.

Mutational studies with the start sequence of the VSV G gene (Stillman and Whitt 1997, 1999) indicated that the first three nucleotides were most critical for efficient gene expression because these nucleotides contained essential signals for processing of the nascent mRNA strand. Without this processing, most of the transcripts were prematurely terminated. The gene ending sequences consist of a tetranucleotide (AUAC in VSV) and the U_7 track (Ross and Whitt 2000). It has been shown with VSV studies that termination of the previous mRNA is affected most if each of the first three nucleotides AUA is replaced with a C residue. As for the 5′ C residue of the tetranucleotide, any replacement resulted in a total loss of termination signaling ability (Barr et al. 1997a; Hwang et al. 1998). The function of the U7 tract has been suggested to provide a template for generation of the mRNA poly (A) tail (Schubert et al. 1980). Mutational analysis of the U tract revealed that shortening or interrupting the U7 tract abolished all termination ability of the resulting gene junction (Barr et al. 1997a). However, increasing the U-tract length had little effect on termination signaling ability. In addition, U7 tract also plays a role in signaling initiation of downstream mRNA synthesis. Either reducing or increasing the length of the U7 tract resulted in reduced downstream initiation (Hinzman et al. 2002). The U7 track has been found to be universal for each of the genes in all the rhabdoviruses (Tordo et al. 1986b; Schutze et al. 1995, 1999; Tanno et al. 2000; Hoffmann et al. 2002). However, the sequences immediately preceding the U7 track may differ from one genus to the other. For example, AC or UC are the ending sequences prior to the U7 track for lyssaviruses (Tordo et al. 1986b; Conzelman et al. 1990; Ito et al. 2001; Warrilow et al. 2002).

Likewise, the intergenic sequences are different from one genus to the other among the rhabdoviruses. However, they are more conserved within each of the genera. For example, the intergenic sequence for VSV is either GA (for N/P, P/M, G/L) or CA (M/G) (Rose 1980). However, the intergenic sequence for rabies is GA (N/P), GUCCG (P/M), GAUAA (M/G), 423 nucleotides (G/L) (Tordo et al. 1986b). The role of these intergenic sequences has been studied with regard to its roles in transcription termination and subsequent downstream transcription in VSV Indiana strain (Stillman and Whitt 1997, 1998; Barr et al. 1997b). Any dinucleotide can signal for termination. However, any mutation on the dinucleotide abolishes the downstream transcription. This finding suggested that the role of the dinucleotide in signaling termination was to position a non-U residue directly downstream of the minimum length of U tract that supported reiterative transcription, while the intergenic sequence acted to physically separate the upstream U tract from the downstream gene start sequence, which begins with the sequence 3′-UUGUC-5′ (Barr et al. 2002). It is thus clear that the gene end sequence, the nontranscribed intergenic sequence, and the gene start sequence are sufficient to direct the polymerase to terminate the upstream mRNA synthesis, to add the polyA tail, and to subsequently initiate the synthesis of downstream mRNA (Schnell at al. 1996).

7.2
Trans-elements Involved in Viral Transcription and Replication

Trans-elements involved in rhabdoviral transcription and replication include viral proteins and possibly some cellular proteins. Cellular proteins, for example, have been found to be packaged in rabies virions (Sagara et al. 1995, 1997) or associated with VSV RdRp (Das et al. 1998; Gupta et al. 2002). These cellular proteins may play a role in the process of viral transcription and replication. For viral proteins, it has been shown conclusively from recent reverse genetics studies that N, P, and L are absolutely required for the initiation of viral transcription and replication (Pattnaik and Wertz 1990; Conzelmann and Schnell 1994; Biacchesi et al. 2000). The minigenome or the full-length genome synthesized by transfected plasmids can only be replicated and subsequently transcribed when N, P, and L are supplied *in trans*. It is not only these viral components per se, but also the complicated interactions between these proteins and viral genomic RNA that initiates and regulates rhabdoviral transcription and replication (Banerjee 1987; Banerjee and Chattopadhyay 1990; Wertz et al. 1987; Wunner 1991). For example, rhabdoviral N encapsidates the de novo-synthesized genomic RNA, thus switching the virus from the mode of transcription to that of replication (Blumberg et al. 1983). However, N is capable of encapsidating nonspecific RNA and the N needs to interact with P for specific encapsidation of the genomic RNA (Banerjee et al. 1989; Yang et al. 1998). Likewise, L interacts with P to form the RdRp to fulfill all the required enzymatic reactions in the process of RNA transcription and replication (Banerjee and Chattopadhyay 1990). Until recently, it was generally believed that the RdRp was both the transcriptase and the replicase. Regulation of the process from transcription to replication is due to the encapsidation of the leader RNA by N, or the N–P complex.

In a recent study, Pattnaik et al. (1997) reported that transcriptionally inactive P mutants can efficiently function in replication of VSV defective interfering particle and these investigators proposed that a tripartite complex consisting of L-(N-P) protein may represent the putative replicase for synthesis of the full-length genome RNA. The tripartite complex has been purified from VSV-infected BHK cells as well as in insect cells that express L, N, and P proteins. The purified tripartite complex supported the replication of genome-sense RNA in an in vitro replication reconstitution reaction (Gupta et al. 2003). In this system, a mutant P protein (P260A) that has been shown to be inactive in transcription but active in replication (Das et al. 1997) was also capable of forming the mutant [L-(N-Pmut)] complex in both insect cells and BHK cells and supporting genome-sense RNA synthesis (Gupta et al. 2003). These studies may indicate that the transcriptase and replicase are structurally distinct entities with different compositions and functions. The transcriptase holoenzyme is composed of L protein bound to phosphorylated P protein oligomer (Gao and Lenard 1995a, b) and specific host factors (Das et al. 1998; Gupta et al. 2002) and is primarily involved in the synthesis of capped mRNAs. The replicase, on the other hand, is composed of L protein bound to the N–P complex, forming a tripartite complex of unknown composition, and it initiates the synthesis of the uncapped leader RNA, which is concomitantly encapsidated by the N protein and continues to synthesize the full-length antigenomic RNA and, subsequently, the genomic RNA. The putative transcriptase and replicase may enter at separate promoter sites on the viral genome (see next section).

7.3 Polymerase Entry site(s) for Transcription and Replication

It has long been thought that rhabdoviral transcription and replication initiate at position 1 of the genome in the VSV system (Blumberg et al. 1981, 1983; Emerson 1982). This hypothesis is formed partially based on in vitro experiments in which purified templates and polymerase were mixed together in the presence of limiting NTPs. In the presence of ATP and CTP, transcription reactions yielded predominantly the dinucleotide 5′ AC 3′, implying that the polymerase only initiated at the leader start because the leader RNA begins 5′ AC... 3′. The first VSV transcript begins with 5′ AACA... 3′ and this tetranucleotide was found to be synthesized in a two-step reaction in which complexes were preincubated in the presence of all four NTPs first. After the reaction was stopped and NTPs removed, the products synthesized in the presence of ATP and CTP were analyzed. These data led to the conclusion that preincubation of the template with all four NTPs allowed polymerase to move along the template, thereby gaining access to the subsequent gene start sequence. Therefore it is hypothesized that polymerase entered the genome at a single site in the leader region and initiated at position 1 of the genome.

While the single-entry site model is favored, evidence has accumulated that is in conflict with this model (see Barr et al. 2002 and references therein). Most notably, experiments performed on a mutant VSV, polR1, which has a single amino acid change in the template-associated N protein, suggested a two-entry site model (Chuang and Perrault 1997; Perrault et al. 1983). PolR1 virus produces a molar excess of N mRNA over the leader RNA in vitro, a finding that was incompatible with transcription of N mRNA requiring transcription through the leader region. This observation prompted Chuang and Perrault (1997) to propose a two-entry site model: 3′ entry for leader synthesis and RNA replication, and direct entry at the N gene start sequence for transcription of the five viral mRNAs. This hypothesis has been confirmed by recent studies with 60- or 108-nucleotide-long genes inserted at the leader-N gene junction (Whelan and Wertz 2002), which suggests that transcription initiated directly at the first gene start sequence and not at the extreme 3′ end of the genome.

8 Conclusions

Rhabdoviruses are one of the few families of viruses whose host range extends from plants, to invertebrates, and to vertebrate animals. Rhabdoviruses not only induce severe diseases in humans and animals, but also cause problems in plants. Most rhabdoviruses are transmitted by insect vectors and these include all the plant-infecting rhabdoviruses, ephemeroviruses, and some vesiculoviruses. Recent data indicate that rhabdoviruses in the Lyssavirus genus can be transmitted by bats, although the classical rabies virus can also be transmitted by carnivores as well. Traditional classification of rhabdoviruses relied on the distinct bullet-shaped morphology and serological relatedness. Genetic analysis and comparison clearly show that these rhabdoviruses are more closely related genetically with each other than to viruses in other families. Sequence analysis has also grouped viruses within some of the genera, for example, lyssaviruses can be grouped into seven genotypes according to genetic relatedness.

With the development of reverse genetics, the functions of many *cis*- and *trans*-elements important in the process of viral transcription and replication are clearly defined, particularly in the VSV system. *cis*-Elements include the 3′ leader and 5′ trailer sequences, gene ending and starting sequences, and the intergenic sequences. The 3′ leader sequences of the genome act as a promoter for both transcription and replication while the 5′ trailer sequence of the genome acts exclusively as an origin of replication. The 3′ end of the antigenome, which is the complementary sequence of the trailer region, acts as the promoter for synthesis of full-length genomic RNA. Gene ending and starting sequences as well as the intergenic sequences have been shown to be important for the proper termination of the preceding transcript and initiation of the subsequent gene. *trans*-Elements include viral proteins N, P, and L, and cellular proteins may also play a role in the process of viral transcription and replication. Recent studies have also challenged conventional wisdom on RdRp complex involved in transcription and replication and the polymerase entry site.

Traditionally, it is believed that the RdRp (L and P) carries out both the RNA transcription and replication, with N-P complex initiating the latter process. Recent data suggest that the transcriptase holoenzyme is composed of L protein bound to phosphorylated P protein oligomer and specific host factors and is primarily involved in the synthesis of capped mRNAs. The replicase, on the other hand, is composed of L protein bound to the N–P complex, forming a tripartite complex of unknown composition, and it initiates the synthesis of the full-length antigenomic RNA and, subsequently, the genomic RNA. Likewise, it was proposed previously that both rhabdoviral transcription and replication initiate at position 1 of the genome. Recent data indicate that there are two entry sites for the RdRp: the 3′ entry for leader synthesis and RNA replication and direct entry at the N gene start sequence for transcription of the monocistronic mRNAs.

Acknowledgements The author expresses gratitude to Mr. Zhi Wei Wang for helping in the analysis of the genetic relatedness of the all the rhabdoviral genes and to the National Institute of Allergy and Infectious Diseases for funding (AI 51560).

References

Abraham G, Banerjee AK (1976) Sequential transcription of the genes of vesicular stomatitis virus. Proc Natl Acad Sci U S A 73:1504–1508

Ahne W, Bjorklund HV, Essbauer S, Fijan N, Kurath G, Winton JR (2002) Spring viremia of carp (SVC). Dis Aquat Organ 52:261–72

Amelia PA, da Rosa T, Mather TN, Takeda T, Whitehouse CA, Shope RE, Popov VL, Gusman H, Coffey L, Araujo TP, Tesh RB (2002) Two new rhabdoviruses (Rhabdoviridae) isolated from birds during surveillance for arboviral encephalitis in Northeastern United States. Emerg Infect Dis 8:614–618

Amengual B, Whitby JE, King A, Cobo JS, Bourhy H (1997) Evolution of European bat lyssaviruses. J Gen Virol 78:2319–28

Arguin PM, Murray-Lillibridge K, Miranda ME, Smith JS, Calaor AB, Rupprecht CE (2002) Serologic evidence of Lyssavirus infections among bats, the Philippines. Emerg Infect Dis 8:258–262

Badrane H, Bahloul C, Perrin P, Tordo N (2001) Evidence of two Lyssavirus phylogroups with distinct pathogenicity and immunogenicity. J Virol 75:3268–3276

Ball LA, White CN (1976) Order of transcription of genes of vesicular stomatitis virus. Proc Natl. Acad Sci U S A 73:442–446

Baltimore D, Huang AS, Stampfer M (1970) Ribonucleic acid synthesis of vesicular stomatitis virus II An RNA polymerase in the virion. Proc Natl Acad Sci U S A 66:572–576

Banerjee AK (1987) Transcription and replication of rhabdoviruses. Microbiol Rev 51:66–87

Banerjee AK, Chattopadhyay D (1990) Structure and function of the RNA polymerase of vesicular stomatitis virus. Adv Virus Res 38:99–124

Banerjee AK, Masters PS, Das T, Chattopadhyay D (1989) Specific interaction of vesicular stomatitis virus nucleocapsid protein (N) with the phosphoprotein (NS) prevents its binding with non-specific RNA. In: Kolakofsky D, Mahy B (eds) Genetics and pathogenicity of negative strand Viruses, Elsevier, Amsterdam, pp 121–128

Barr J, Chambers P, Pringle CR, Easton AJ (1991) Sequence of the major nucleocapsid protein gene of pneumonia virus of mice: sequence comparisons suggest structural homology between nucleocapsid proteins of pneumoviruses, paramyxoviruses, rhabdoviruses and filoviruses. J Gen Virol 72:677–685

Barr JN, Whelan SP, Wertz GW (1997a) *cis*-Acting signals involved in termination of vesicular stomatitis virus mRNA synthesis include the conserved AUAC and the U7 signal for polyadenylation. J Virol 71:8718–8725

Barr JN, Whelan SP, Wertz GW (1997b) Role of the intergenic dinucleotide in vesicular stomatitis virus RNA transcription. J Virol 71:1794–1801

Barr JN, Whelan SP, Wertz GW (2002) Transcriptional control of the RNA-dependent RNA polymerase of vesicular stomatitis virus. Biochim Biophys Acta 1577:337–353

Biacchesi S, Yu YX, Bearzotti M, Tafalla C, Fernandez-Alonso M, Bremont M (2000) Rescue of synthetic salmonid rhabdovirus minigenomes. J Gen Virol 81:1941–1945

Blumberg BM, Leppert M, Kolakofsky D (1981) Interaction of VSV leader RNA and nucleocapsid protein may control VSV genome replication. Cell 23:837–45

Blumberg BM, Giorgi C, Kolakofsky D (1983) N protein of vesicular stomatitis virus selectively encapsidates leader RNA in vitro. Cell 32:559–567

Bourhy H, Kissi B, Tordo N (1993) Molecular diversity of the Lyssavirus genus. Virology 194:70–81

Brown F (1987) The family Rhabdovidae: general description and taxonomy. In: Wagner RR (ed) The rhabdoviruses. Plenum Press, New York, pp 1–8

Calisher CH, Karabatsos N, Dalrymple JM, Shope RE, Porterfield JS, Westaway EG, Brandt WE (1989) Antigenic relationships between flaviviruses as determined by cross-neutralization tests with polyclonal antisera. J Gen Virol 70:37–43

Chandrika R, Horikami SM, Smallwood S, Moyer SA (1995) Mutations in conserved domain I of the Sendai virus L polymerase protein uncouple transcription and replication. Virology 213:352–363

Chen XY, Luo ZL, Fang RX (1998) Structure analysis of the rice yellow stunt rhabdovirus gene 3 (in Chinese). Chin Sci Bull 43:745–748

Chenik M, Chebli K, Blondel D (1995) Translation initiation at alternate in-frame AUG codons in the rabies virus phosphoprotein mRNA is mediated by a ribosomal leaky scanning mechanism. J Virol 69:707–712

Chuang JL, Perrault J (1977) Initiation of vesicular stomatitis virus mutant polR1 transcription internally at the N gene in vitro. J Virol 71:1466–1475

Colonno RJ, Banerjee AK (1978) Complete nucleotide sequence of the leader RNA synthesized in vitro by vesicular stomatitis virus. Cell 15:93–101

Conzelmann KK, Cox JH, Schneider LG, Thiel HJ (1990) Molecular cloning and complete nucleotide sequence of the attenuated rabies virus SAD B19. Virology 175:485–499

Das T, Pattnaik AK, Takacs AM, Li T, Hwang LN, Banerjee AK (1997) Basic amino acid residues at the carboxy-terminal eleven amino acid region of the phosphoprotein (P) are required for transcription but not for replication of vesicular stomatitis virus genome RNA. Virology 238:103–114

Das T, Mathur M, Gupta AK, Janssen GMC, Banerjee AK (1998) RNA polymerase of vesicular stomatitis virus specifically associates with translation elongation factor-1$\alpha\beta\gamma$ for its activity. Proc Natl Acad Sci U S A 95:1449–1454
Diaz AM, Papo S, Rodriguez A, Smith JS (1994) Antigenic analysis of rabies-virus isolates from Latin America and the Caribbean. J Vet Med B 41:153–160
Dietzschold B, Rupprecht CE, Fu ZF, Koprowski H (1996) Rhabdoviruses. In: Fields B, Knipe D, Howley PM et al (eds) Fields' Virology, 3rd edn. Lippincott-Reven, Philadelphia, pp 1137–1159
Emerson SU (1982) Reconstitution studies detect a single polymerase entry site on the vesicular stomatitis virus genome. Cell 31: 635–642
Emerson SU, Yu Y (1975) Both NS and L proteins are required for in vitro RNA synthesis by vesicular stomatitis virus. J Virol 15:1348–1356
Finke S, Conzelmann KK (2003) Dissociation of rabies virus matrix protein functions in regulation of viral RNA synthesis and virus assembly. J Virol 77:12074–1282
Fu ZF (1997) Rabies and rabies research: past, present and future. Vaccine 15:S20–S24
Gao Y, Lenard J (1995a) Cooperative binding of multimeric phosphoprotein (P) of vesicular stomatitis virus to polymerase (L) and template. J Virol 69:7718–7723
Gao Y, Lenard J (1995b) Multimerization and transcriptional activation of the phosphoprotein (P) of vesicular stomatitis virus by casein kinase II. EMBO J 14:1240–1247
Gallione CJ, Rose JK (1983) Nucleotide sequence of a cDNA clone encoding the entire glycoprotein from the New Jersey serotype of vesicular stomatitis virus. J Virol 46:162–169
Gillies S, Stollar V (1980) The production of high yields of infectious vesicular stomatitis virus in A albopictus cells and comparisons with replication in BHK-21 cells. Virology 107:509–513
Gould AR, Hyatt AD, Lunt R, Kattenbelt JA, Hengstberger S, Blacksell SD (1998) Characterisation of a novel lyssavirus isolated from pteropid bats in Australia. Virus Res 54:165–187
Gould AR, Kattenbelt JA, Gumley SG, Lunt RA (2002) Characterisation of an Australian bat lyssavirus variant isolated from an insectivorous bat. Virus Res 89:1–28
Green TJ, Macpherson S, Qiu S, Lebowitz J, Wertz GW, Luo M (2000) Study of the assembly of vesicular stomatitis virus N protein: role of the P protein. J Virol 74:9515–9524
Gupta AK, Mathur M, Banerjee AK (2002) Unique capping activity of the recombinant RNA polymerase (L) of vesicular stomatitis virus: association of cellular capping enzyme with the L protein. Biochem Biophys Res Commun 293:264–268
Gupta AK, Shaji D, Banerjee AK (2003) Identification of a novel tripartite complex involved in replication of vesicular stomatitis virus genome RNA. J Virol 77:732–738
Herman RC (1986) Internal initiation of translation on the vesicular stomatitis virus phosphoprotein mRNA yields a second protein. J Virol 58:797–804
Hinzman E, Barr JN, Wertz GW (2002) Identification of an upstream sequence element required for vesicular stomatitis mRNA transcription. J Virol 76:7632–7641
Hoffmann B, Schutze H, Mettenleiter TC (2002) Determination of the complete genomic sequence and analysis of the gene products of the virus of Spring Viremia of Carp, a fish rhabdovirus. Virus Res 84:89–100
Hogenhout SA, Redinbaugh MG, Ammar ED (2003) Plant and animal rhabdovirus host range: a bug's view. Trends Microbiol 11:264–271

Huang Y, Zhao H, Luo Z, Chen X, Fang R (2003) Novel structure of the genome of Rice yellow stunt virus: identification of the gene 6-encoded virion protein. J Gen Virol 84: 2259–2264

Hwang LN, Englund N, Pattnaik AK (1998) Polyadenylation of vesicular stomatitis virus mRNA dictates efficient transcription termination at the intercistronic gene junctions. J Virol 72:1805–1813

Ito N, Kakemizu M, Ito KA, Yamamoto A, Yoshida Y, Sugiyama M, Minamoto N (2001) A comparison of complete genome sequences of the attenuated RC-HL strain of rabies virus used for production of animal vaccine in Japan, and the parental Nishigahara strain. Microbiol Immunol 45:51–58

Iverson LE, Rose JK (1981) Localized attenuation and discontinuous synthesis during vesicular stomatitis virus transcription. Cell 23:477–484

Jackson AO, Francki RIB, Zuidema D (1987) Biology, structure, and replication of plant rhabdoviruses. In: Wagner RR (ed) The rhabdoviruses. Plenum Press, New York, pp 427–508

Jonkers AH, Shope RE, Aitken TH, Spence L (1964) Cocal virus: a new agent In Trinidad related to vesicular stomatitis virus type Indiana. Am J Vet Res 25:236–242

Krebs JW, Smith JS, Rupprecht CE, Childs JE (2000) Mammalian reservoirs and epidemiology of rabies diagnosed in human beings in the United States, 1981–1998. Ann N Y Acad Sci 916:345–353

Kurath G, Ahern KG, Pearson GD, Leong JC (1985) Molecular cloning of the six mRNA species of infectious hematopoietic necrosis virus, a fish rhabdovirus, and gene order determination by R-loop mapping. J Virol 53:469–476

Kurath G, Higman KH, Bjorklund HV (1997) Distribution and variation of NV genes in fish rhabdoviruses. J Gen Virol 78:113–117

Kuzmin IV, Orciari LA, Arai YT, Smith JS, Hanlon CA, Kameoka Y, Rupprecht CE (2003) Bat lyssaviruses (Aravan and Khujand) from Central Asia: phylogenetic relationships according to N, P and G gene sequences. Virus Res 97:65–79

Landes-Devauchelle C, Bras F, Dezelee S, Teninges D (1995) Gene 2 of the sigma rhabdovirus genome encodes the P protein, and gene 3 encodes a protein related to the reverse transcriptase of retroelements. Virology 213:300–312

Larson JK, Wunner WH (1990) Nucleotide and deduced amino acid sequences of the nominal nonstructural phosphoprotein of the ERA, PM and CVS-11 strains of rabies virus. Nucleic Acids Res 18:7172

Li T, Pattnaik AK (1999) Overlapping signals for transcription and replication at the 3′ terminus of the vesicular stomatitis virus genome. J Virol 73:444–452

Martins CR, Johnson JA, Lawrence DM, Choi TJ, Pisi AM, Tobin SL, Lapidus D, Wagner JD, Ruzin S, McDonald K, Jackson AO (1998) Sonchus yellow net rhabdovirus nuclear viroplasms contain polymerase-associated proteins. J Virol 72:5669–5679

McWilliam SM, Kongsuwan K, Cowley JA, Byrne KA, Walker PJ (1997) Genome organization and transcription strategy in the complex G_{NS}-L intergenic region of bovine ephemeral fever rhabdovirus. J Gen Virol 78, 1309–1317

Messenger SL, Smith JS, Rupprecht CE (2002) Emerging epidemiology of bat-associated cryptic cases of rabies in humans in the United States. Clin Infect Dis 35:738–747

Morimoto K, Patel M, Corisdeo S, Hooper DC, Fu ZF, Rupprecht CE, Koprowski H, Dietzschold B (1996) Characterization of a unique variant of bat rabies virus responsible for newly emerging human cases in North America. Proc Natl Acad Sci U S A 93:5653–5658

Mork C, Hershberger P, Kocan R, Batts W, Winton J (2004) Isolation and characterization of a rhabdovirus from starry flounder (Platichthys stellatus) collected from the northern portion of Puget Sound Washington USA. J Gen Virol 85:495–505
Muller R, Poch O, Delarue M, Bishop DH, Bouloy M (1994) Rift Valley fever virus L segment: correction of the sequence and possible functional role of newly identified regions conserved in RNA-dependent polymerases. J Gen Virol 75:1345–1352
Murphy (1996) Virus taxonomy. In: Fields B, Knipe D, Howley PM et al (eds) Fields' Virology, 3rd edn. Lippincott-Reven, Philadelphia, pp 15–57
Nault LR (1997) Arthropod transmission of plant viruses: a new synthesis. Ann Entomol Soc Am 90:521–541
Pattnaik AK, Wertz GW (1990) Replication and amplification of defective interfering particle RNAs of vesicular stomatitis virus in cells expressing viral proteins from vectors containing cloned cDNAs. J Virol 64:2948–57
Pattnaik AK, Wertz GW (1991) Cells that express all five proteins of vesicular stomatitis virus from cloned cDNAs support replication, assembly, and budding of defective interfering particles. Proc Natl Acad Sci U S A 88:1379–1383
Pattnaik AK, Ball LA, LeGrone A, Wertz GW (1995) The termini of VSV DI particle RNAs are sufficient to signal RNA encapsidation, replication, and budding to generate infectious particles. Virology 206:760–764
Pattnaik AK, Hwang L, Li T, Englund N, Mathur M, Das T, Banerjee AK (1997) Phosphorylation within the amino-terminal acidic domain I of the phosphoprotein of vesicular stomatitis virus is required for transcription but not for replication. J Virol 71:8167–8175
Perrault J, Clinton GM, McClure MA (1983) RNP template of vesicular stomatitis virus regulates transcription and replication functions. Cell 35:175–185
Poch O, Sauvaget I, Delarue M, Tordo N (1989) Identification of four conserved motifs among the RNA-dependent polymerase encoding elements. EMBO J 8:3867–3874
Poch O, Blumberg BM, Bougueleret L, Tordo N (1990) Sequence comparison of five polymerases (L proteins) of unsegmented negative-strand RNA viruses: theoretical assignment of functional domains. J Gen Virol 71:1153–1162
Printz P (1973) Relationship of Sigma virus to vesicular stomatitis virus. Adv Virus Res 18:143
Rayssiguier C, Cioe L, Withers E, Wunner WH, Curtis PJ (1986) Cloning of rabies virus matrix protein mRNA and determination of its amino acid sequence. Virus Res 5:177–190
Rose JK (1980) Complete intergenic and flanking gene sequences from the genome of vesicular stomatitis virus. Cell 19:415–421
Rose JK, Whitt MA (2000) Rhabdoviridae: the virus and their replication. In: Knipe DM, Howley PM, Griffin DE, Lamb RA, Martin MA, Roizman B, Straus SE (eds) Field's Virology, 4th edn. Lippincott-Reven, Philadelphia, pp 1221–1240
Sagara J, Tsukita S, Yonemura S, Tsukita S, Kawai A (1995) Cellular actin-binding ezrin-radixin-moesin (ERM) family proteins are incorporated into the rabies virion and closely associated with viral envelope proteins in the cell. Virology 206:485–494
Sagara J, Tochikura TS, Yamamoto T, Tsukita S, Tsukita S, Kawai A (1997) Immunological studies of a 21 kDa cellular component efficiently incorporated into rabies virion grown in a BHK-21 cell culture. Microbiol Immunol 41:947–955
Saitou N, Nei M (1987) The neighbor-joining method: a new method for reconstructing phylogenetic trees. Mol Biol Evol 4:406–425
Schnell MJ, Conzelmann KK (1995) Polymerase activity of in vitro mutated rabies virus L protein. Virology 214:522–530

Schnell MJ, Mebatsion T, Conzelmann KK (1994) Infectious rabies viruses from cloned cDNA. EMBO J 13:4195–4203
Schnell MJ, Buonocore L, Whitt MA, Rose JK (1996) The minimal conserved transcription stop-start signal promotes stable expression of a foreign gene in vesicular stomatitis virus. J Virol 70:2318–2323
Schoehn G, Iseni F, Mavrakis M, Blondel D, Ruigrok RW (2001) Structure of recombinant rabies virus nucleoprotein-RNA complex and identification of the phosphoprotein binding site. J Virol 75:490–498
Schubert M, Keene JD, Herman RC, Lazzarini RA (1980) Site on the vesicular stomatitis virus genome specifying polyadenylation and the end of the L gene mRNA J Virol 34:550–559
Schubert M, Harmison GG, Richardson CD, Meier E (1985) Expression of a cDNA encoding a functional 241-kilodalton vesicular stomatitis virus RNA polymerase. Proc Natl Acad Sci U S A 82:7984–7988
Schutze H, Enzmann PJ, Kuchling R, Mundt E, Niemann H, Mettenleiter TC (1995) Complete genomic sequence of the fish rhabdovirus infectious haematopoietic necrosis virus. J Gen Virol 76:2519–2527
Schutze H, Mundt E, Mettenleiter TC (1999) Complete genomic sequence of viral hemorrhagic septicemia virus, a fish rhabdovirus. Virus Genes 19:59–65
Shope RE (1982) Rabies-related viruses. Yale J Biol Med 55:271–275
Shope RE, Tesh RB (1987) The ecology of rhabdoviruses that infect vertebrates. In: Wagner RR (ed) The rhabdoviruses. Plenum Press, New York, pp 509–534
Smallwood S, Easson CD, Feller JA, Horikami SM, Moyer SA (1999) Mutations in conserved domain II of the large (L) subunit of the Sendai virus RNA polymerase abolish RNA synthesis. Virology 262:375–383
Spiropoulou CF, Nichol ST (1993) A small highly basic protein is encoded in overlapping frame within the P gene of vesicular stomatitis virus. J Virol 67:3103–3110
St George TD (1990). Bovine ephemeral fever virus. In: Dinter Z and Morein B (eds) Virus infections of vertebrates. Vol 3: Virus infections of ruminants. Elsevier, Amsterdam, pp 405–415
Stallknecht DE (2000) VSV-NJ on Ossabaw Island, Georgia. The truth is out there. Ann N Y Acad Sci 916:431–436
Stillman EA, Whitt MA (1997) Mutational analyses of the intergenic dinucleotide and the transcriptional start sequence of vesicular stomatitis virus (VSV) define sequences required for efficient termination and initiation of VSV transcripts. J Virol 71:2127–2137
Stillman EA, Whitt MA (1998) The length and sequence composition of vesicular stomatitis virus intergenic regions affect mRNA levels and the site of transcript initiation. J Virol 72:5565–5572
Stillman EA, Whitt MA (1999) Transcript initiation and 5′-end modifications are separable events during vesicular stomatitis virus transcription. J Virol 73:7199–7209
Tanno F, Nakatsu A, Toriyama S, Kojima M (2000) Complete nucleotide sequence of Northern cereal mosaic virus and its genome organization 145:1373–1384
Teninges D, Bras F, Dezelee S (1993) Genome organization of the sigma rhabdovirus: six genes and a gene overlap. Virology 193:1018–1023
Tordo N, Poch O, Ermine A, Keith G (1986a) Primary structure of leader RNA and nucleoprotein genes of the rabies genome: segmented homology with VSV. Nucleic Acids Res 14:2671–2683

Tordo N, Poch O, Ermine A, Keith G, Rougeon F (1986b) Walking along the rabies genome: is the large G-L intergenic region a remnant gene? Proc Natl Acad Sci U S A 83:3914–3918
Tordo N, Poch O, Ermine A, Keith G, Rougeon F (1988) Completion of the rabies virus genome sequence determination: highly conserved domains among the L (polymerase) proteins of unsegmented negative-strand RNA viruses. Virology 165:565–576
Villarreal LP, Breindl M, Holland JJ (1976) Determination of molar ratios of vesicular stomatitis virus induced RNA species in BHK21 cells. Biochemistry 15:1663–1667
Wagner RR (1987) Rhabdovirus biology and infection: an overview. In: Wagner RR (ed) The rhabdoviruses. Plenum Press, New York, pp 9–74
Walker PJ, Byrne KA, Riding GA, Cowley JA, Wang Y, McWilliam S (1992). The genome of bovine ephemeral fever rhabdovirus contains two related glycoprotein genes. Virology 191:49–61
Walker PJ, Wang Y, Cowley JA, McWilliam SM, Prehaud CJ (1994) Structural and antigenic analysis of the nucleoprotein of bovine ephemeral fever rhabdovirus. J Gen Virol 75:1889–1899
Walker PJ, Benmansour A, Dietzgen R, Fang RX, Jackson AO, Kurath G, Leong JC, Nadin-Davis S, Tesh RB, Tordo N (2000) Rhabdoviridae. In: Van Regenmortel MHV, Fauquet CM, Bishop DHL, Carstens EB, Estes MK, Lemon SM, Miniloff J, Mayo MA, McGeoch DJ, Pringle CR, Wickner RB (eds) Virus taxonomy. Seventh report of the International Committee on Taxonomy of Viruses. Academic Press, New York, pp 562–583
Wang Y, Cowley JA, Walker PJ (1995) Adelaide River virus nucleoprotein gene: analysis of phylogenetic relationships of ephemeroviruses and other rhabdoviruses. J Gen Virol 76:995–999
Wang Y, McWilliam SM, Cowley JA, Walker PJ (1994) Complex genome organization in the G_{NS}-L intergenic region of Adelaide River rhabdovirus. Virology 203:63–72
Warrilow D, Smith IL, Harrower B, Smith GA (2002) Sequence analysis of an isolate from a fatal human infection of Australian bat lyssavirus. Virology 297:109–119
Wertz GW, Davies NL, Patton J (1987) The role of proteins in vesicular stomatitis virus RNA replication. In: Wagner RR (ed) The rhabdoviruses. Plenum Press, New York, pp 271–296
Wertz GW, Whelan S, LeGrone A, Ball LA (1994) Extent of terminal complementarity modulates the balance between transcription and replication of vesicular stomatitis virus RNA. Proc Natl Acad Sci U S A 91:8587–8591
Wetzel T, Dietzgen RG, Dale JL (1994) Genomic organization of lettuce necrotic yellows rhabdovirus. Virology 200:401–412
Whelan SP, Wertz GW (1999) Regulation of RNA synthesis by the genomic termini of vesicular stomatitis virus: identification of distinct sequences essential for transcription but not replication. J Virol 73:297–306
Whelan SP, Wertz GW (2002) Transcription and replication initiate at separate sites on the vesicular stomatitis virus genome. Proc Natl Acad Sci U S A 99:9178–9183
Wunner WH (1991) The chemical composition and molecular structure of rabies viruses. In Baer GM (ed), Natural history of rabies, 2nd edn. CRC Press, Boca Raton, pp 31–67
Yang J, Hooper DC, Wunner WH, Koprowski H, Dietzschold B, Fu ZF (1998) The specificity of rabies virus RNA encapsidation by nucleoprotein. Virology 242:107–117

CTMI (2005) 292:25–44

Australian Bat Lyssavirus: A Recently Discovered New Rhabdovirus

D. Warrilow

Public Health Virology Laboratory, Queensland Health Scientific Services, 39 Kessels Rd , 4108 Coopers Plains, Queensland, Australia
David.Warrilow@qimr.edu.au

Abstract Australian bat lyssavirus (ABLV), first identified in 1996, has been associated with two human fatalities. ABLV is genetically and serologically distinct from, but is closely related to, classical rabies. It has a bullet-shaped morphology by electron microscopy. There are two strains of ABLV known: one circulates in frugivorous bats, sub-order *Megachiroptera*, and the other circulates in the smaller, mainly insectivorous bats, sub-order *Microchiroptera*. Each strain has been associated with one human fatality. Surveillance indicates infected bats are widespread at a low frequency on the Australian mainland. It is unclear how long ABLV has been present in Australia, although molecular clock studies suggest the two strains separated 950 or 1,700 years ago based on synonymous or non-synonymous nucleotide changes, respectively. Recent serological surveys suggest a closely related virus may exist in the Philippines. Due to demonstrated cross-protection in mice, rabies vaccine is used to prevent infection. Rabies post-exposure prophylaxis (PEP) protocols have been adopted for when a human is scratched or bitten by a suspect bat. A long-term commitment to public health programs that test bats that have been involved in scratch or bite incidents, followed by PEP if appropriate, will be necessary to minimise further human infection.

1 Introduction

For most of its history, Australia prided itself on being free of rabies and rabies-related viruses. This was a privileged position it shared with only one other continent, Antarctica, and several smaller island states. However, in 1996 it was discovered that Australian bat lyssavirus (ABLV) was present in bat populations in the mainland state of New South Wales. While genetically distinct, ABLV is closely related to classical rabies virus, and its distribution extends from Australia probably into Asia. ABLV is currently circulating as two distinct strains known as *Pteropus* and *Saccolaimus*, in reference to the bat species from which it was originally isolated. Due to its relatively recent identification, there is a paucity of published data available for ABLV, and at this stage it remains unclear whether it has recently emerged in Australian bat populations, or if it has been a long-term feature of the ecology. The objective of this chapter is to provide a broad overview of the available information on the history, pathology, biology and molecular biology of ABLV.

Classical rabies, which has been and still remains a horrific affliction across much of the globe, is not endemic to Australia. Considering its near-ubiquitous distribution, it is interesting that Australia escaped establishment of rabies. Australia's geographical isolation may have protected it from rabies incursions. If rabies was present at some time in the country's distant past, its unique terrestrial biota was incapable of sustaining a permanent infectious cycle up until contemporary time. Only three documented human cases of rabies have been reported in Australia. The first was in Tasmania in 1867 and involved both a dog and a boy (Bisseru 1972). A second case, the first to receive laboratory confirmation, occurred in Queensland in 1987 (Faoagali et al. 1988). In that case the patient, a 10-year-old boy, was most likely infected by the bite of a monkey 16 months prior while on holiday in India. A third case was diagnosed in New South Wales in 1990 in a 10-year-old girl (Grattan-Smith et al. 1992). While no bite incident could be recalled in that case, infection was most likely acquired at least 6 years earlier in Vietnam, her home country. Sequence analysis confirmed this hypothesis (McColl et al. 1993).

2
The Discovery of Australian Bat Lyssavirus

The discovery of Hendra virus, originally referred to as equine morbillivirus, indirectly led to the discovery of ABLV. Two human fatalities were caused by Hendra virus infection from horses (Hooper et al. 1996; Murray et al. 1995a; Murray et al. 1995b; O'Sullivan et al. 1997; Rogers et al. 1996; Selvey et al. 1995). In 1996, a retrospective study was initiated to search for the natural host of Hendra, and pteropid bats (flying foxes) were later identified as the likely reservoir for the virus (Halpin et al. 2000; Young et al. 1996). Whilst doing so, immuno-histochemistry was performed on brain sections from two black flying foxes (*Pteropus alecto*) collected from Ballina, northern New South Wales, which were negative for antibodies to Hendra (Fraser et al. 1996). One sample, collected in 1996, was from a female which was unable to fly, and the other was a paraffin-embedded section collected the previous year from a female that had been acting aggressively. An anti-rabies monoclonal antibody reacted with both specimens. No neutralizing antibodies were detected and virus could not be directly isolated by culture from the bat collected in 1996. Samples from both bats resulted in a product when amplified by reverse transcriptase-polymerase chain reaction (RT-PCR) using degenerate oligonucleotides complementary to the Lyssavirus nucleocapsid protein. Analysis of the products indicated that the sequence of the amplicons from both bats was identical, and similar to the Lyssavirus group. Subsequent intracerebral inoculation of kidney homogenate into weanling mice and further passage in mouse neuroblastoma cells allowed a virus to be isolated, which was later found to be neutralized by antisera to rabies virus (RABV). Monoclonal antibody reactivity of the isolate also revealed a distinctly separate but similar profile to that of the other lyssaviruses.

3
Human Infections

Perhaps co-incidentally, the first human case (case 1) of ABLV infection occurred in October 1996, shortly after its identification in flying foxes (Allworth et al. 1996). A 39-year-old woman presented with pain and numbness in her left arm followed by symptoms including dizziness, headache, fever and vomiting over the next 2–3 days. Her condition deteriorated after hospital admission and she showed signs of diplopia and bulbar palsy. Treatment with acyclovir and antibiotics proved unsuccessful. RT-PCR with ABLV-specific oligonucleotides performed on a cerebrospinal fluid sample resulted in an amplification product. Treatment with rabies immunoglobulin failed and the patient died after 20 days of illness. Of note was the fact the woman had recently been caring for flying foxes and had received a number of scratches to her left arm. At the time, an epidemiological link was made with infected flying foxes on this basis. However, later sequence analysis of the amplification product (Gould et al. 2002) implicated ABLV infection by a bite from an infected yellow-bellied sheath-tailed bat (*Saccolaimus flaviventris*), a strain of which had just been identified prior to this case. Further investigation revealed that the woman had also been bitten by a "microbat" in her care (Tidemann et al. 1997) (the term "microbat" is commonly used to describe the smaller generally insectivorous *Microchiroptera* of which the yellow-bellied sheath-tailed bat is a member).

A second human case (case 2) followed in November 1998. A 37-year-old woman presented with 5 days of fever, vomiting, anorexia, pain in the left shoulder girdle and paraesthesiae of the left hand, sore throat and difficulty swallowing. Her condition deteriorated quickly and she experienced agitation, spasms, dysphagia and dysphonia. She became paralysed and required ventilator support. A diagnosis of ABLV infection was considered and a saliva sample was tested by hemi-nested RT-PCR (Heaton et al. 1997). Subsequent sequencing of the amplicon confirmed an infection with ABLV. On day 14, the patient ceased respiratory effort and she died on day 19. Diagnosis was confirmed post-mortem by immunofluorescent antibody staining and PCR on brain tissue. The time of potential exposure in this case indicates an unusually long incubation period. The patient was involved in an incident 27 months earlier in which she was bitten trying to remove a flying fox that had landed on a child during a barbecue.

A number of attempts have been made to link undiagnosed encephalitis in hospital records with ABLV, but no evidence to support this has been reported (Gerrard 1997; Jong 1997; Lambert et al. 1997; Skull et al. 1999). The above human cases are the only documented naturally acquired ABLV infections in any animals other than bats. There is one unpublished report of transient illness and seroconversion when cats and dogs were experimentally infected (Mackenzie et al. 2003).

4 Flying Fox Infection

Two observations of natural infections provide insight into the clinical course of the ABLV infection in flying foxes. In the first, a male juvenile flying fox, approximately 2–3 weeks of age, developed symptoms after an incubation period of 6–9 weeks and died following 9 days of illness (Field et al. 1999). In the second incident, a black flying fox (*P. alecto*) was found acting aggressively outside the cage of a colony of captive flying foxes (Warrilow et al. 2003). The bat was removed and euthanased, and was found to be positive for ABLV infection. Twenty-nine days later a grey-headed flying fox (*P. poliocephalus*) from the colony started to display signs of a neurological illness. It was euthanased and was also positive for ABLV infection. Circumstantial evidence for an epidemiological link between the two infected bats was made on the basis of a common sequence from the variable 3′ non-coding region of the glycoprotein.

In the only reported study of laboratory infection of flying foxes, three out of ten flying foxes inoculated intra-muscularly with ABLV contracted symptoms after 15, 23 and 24 days (McColl et al. 2002). As the bats were euthanased 1 day after showing signs of illness, the length of illness was not determined. None of the symptomatic bats seroconverted; however, five of seven of the surviving flying foxes produced neutralizing antibodies.

5 Neurology and Pathology in Bats and Humans

ABLV has caused non-suppurative meningoencephalomyelitis in both bats and humans. When histology was performed on bats infected in the wild, eosinophilic cytoplasmic inclusions, known as Negri bodies typical of RABV infection, were often observed (Fraser et al. 1996; Gould et al. 1998; Hooper et al. 1999; Skerratt et al. 1998; Speare et al. 1997). Although no gross lesions in bats have been reported, the following have been described: neuronal necrosis with some neuronophagia, focal gliosis, perivascular cuffing with lymphocytes, cytoplasmic vacuolization and subarachnoid lymphocytic meningitis (Hooper et al. 1999). The medulla oblongata and pons, hippocampus, thalamus and midbrain were most affected. Staining using an immunoperoxidase-labeled rabies-specific monoclonal antibody, HAM, is frequently observed throughout the brain and in particular the hippocampal region, medulla oblongata and pons, thalamus and midbrain, although less commonly in the cerebrum and cerebellum (Fraser et al. 1996; Hooper et al. 1999). Lesions and immunostaining have also been seen in the spinal cord and adjacent dorsal root ganglion of one bat, as well the adrenal medulla, stomach, intestine and aortic tissue in some bats (Hooper et al. 1999). Virus has been detected in one of eight salivary glands of wild bats by immunostaining (Hooper et al. 1997, 1999). When infected experimentally with ABLV, lesions and immunostaining were seen in the brain and spinal cord. Virus was detected in the salivary gland of one laboratory-infected flying fox by RT-PCR (McColl et al. 2002).

There was a degree of variability in the neuropathology of the two human cases. Reduced numbers of neurons were observed in the brain stem, hippocampus and the Purkinje cells of the cerebellum were observed in case 1 (Samaratunga et al. 1998). This was particularly so for pyramidal cells of the hippocampus. Other histological changes included microglial activation, astrocytosis and lymphocyte infiltration, perivascular cuffing and neuronophagia. Eosinophilic inclusions in neurons were observed but these were smaller, up to 7 μm in size, than reported for Negri bodies. These inclusions were reactive with rabies monoclonal antibody. Acinar cells had inclusions and chronic sialadenitis was apparent. Examination by light microscopy revealed that case 2 had a more severe pathology (Hanna et al. 2000). With the exception of the cerebellum, all areas of the brain were affected. The hippocampus and brain stem were particularly necrotic. Perivascular cuffing, infiltration by macrophages and microglial cells, neuronophagia and eosinophilic inclusions were also observed. Brain stem, spinal cord, midbrain, cerebellum and both cerebral hemispheres were positive by immunofluorescent antibody staining.

6 Microscopy and Particle Morphology

Inclusion bodies containing ribonucleoprotein (RNP) were observed by electron microscopy (Fig. 1A) of infected bat, mouse and human brain, and infected cultured cells (Gould et al. 1998; Hooper et al. 1997; Samaratunga et al. 1998). These were composed of crescent-shaped tri-laminar membranes and associated with electron-dense material and RNP filaments. In infected mouse brain, these membranes were also associated with an electron-dense granular material. In addition, ring-shaped inclusions were observed in the dendritic processes or cytoplasm of large neurons, or associated with inclusions.

Fig. 1A, B Transmission electron micrographs of an ABLV-infected BHK-21 cells. **A** Low magnification of an infected BHK-21 cell showing the presence of large inclusion bodies (*); released virus (*arrows*) and an intracellular vesicle containing viruses (*arrowhead*). *Nu*, nucleus; bar represents 1 μm. **B** Viruses at the periphery of the infected BHK-21 cell. Cellular material (*arrow*) and free lying RNP (*open arrow*) are indicated. The presence of cellular debris in association with a viable cell suggests that viruses may also be released by fusion of cytoplasmic-membrane compartments with the plasma membrane. An electron-dense region of the plasma membrane is indicated (*arrowheads*). Bar represents 100 nm. (Reprinted from Virus Research, vol. 54, Gould et al., Characterisation of a novel lyssavirus isolated from Pteropid bats in Australia p. 170, 1998, with permission from Elsevier) ▶

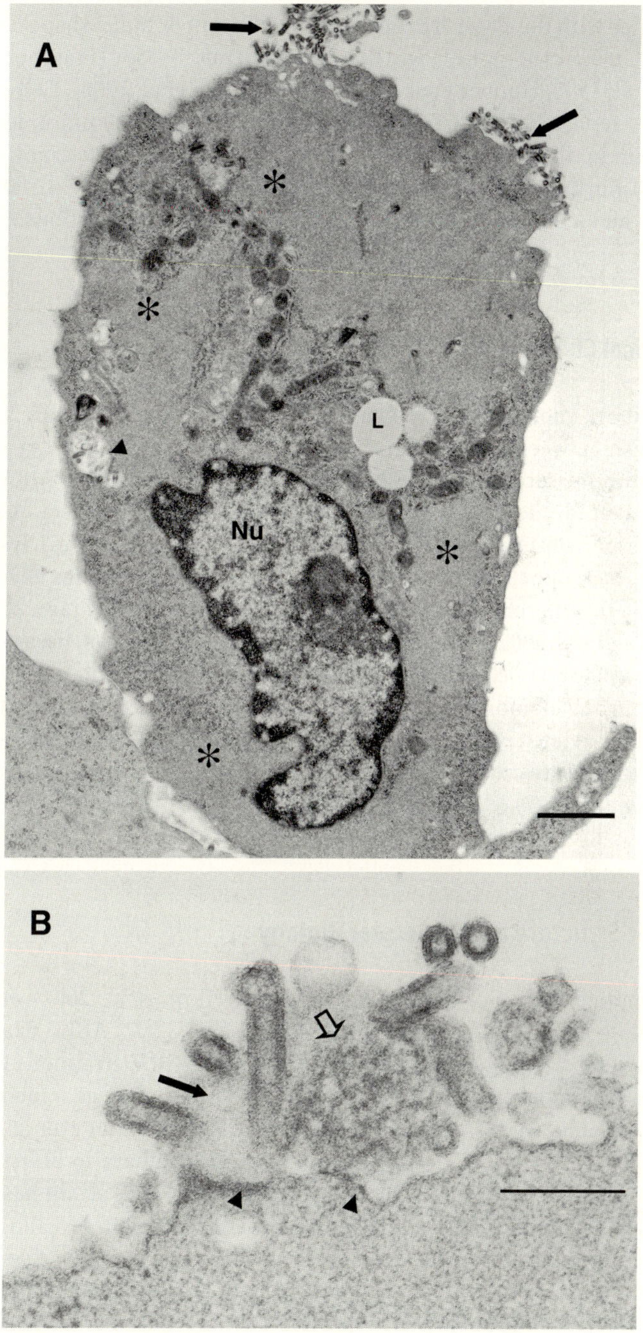
A
L
Nu
B

Particles with the classic rhabdovirus bacilliform or bullet shape have been observed in infected tissue (Fig. 1B) by electron microscopy (Fraser et al. 1996; Gould et al. 1998; Hooper et al. 1997; Samaratunga et al. 1998). Cell-cultured virus was reported to be 65 or 80 nm in diameter and 200 nm in length, or 90–110 nm in diameter and 200 or 250 nm in length in infected mouse brain; with a 40-nm core (Gould et al. 1998; Hooper et al. 1997). The surface of the particles had a honeycombed appearance with 7–9 nm projections.

7
Serological Characteristics

The ability of rabies immune serum to neutralize virus infection is used to distinguish between rabies and rabies-related viruses (Schneider et al. 1973). Rabies immune sera neutralizes both *Pteropus* and *Saccolaimus* ABLV strains using the rapid fluorescent focus inhibition test (RFFIT) (Fraser et al. 1996; Hanlon et al. 2001). Hence, by definition, ABLV is a serotype 1 lyssavirus. However, by monoclonal antibody reactivity, both ABLV strains have similar but distinctly different serological profiles to classical rabies (Fraser et al. 1996; Gould et al. 1988, 2002; Hooper et al. 1997). Using a Centers for Disease Control (CDC) panel of nucleoprotein reactive monoclonal antibodies, the *Pteropus* strain of ABLV demonstrated greater similarity to classical rabies than the other rabies-related lyssaviruses (Hooper et al. 1997). Commercially available rabies vaccines are cross-protective against ABLV infection in a mouse model (Hooper et al. 1997).

8
Genomic Structure and Molecular Biology

The complete sequence of the *Pteropus* (Warrilow et al. 2002) and near-complete sequence of the *Saccolaimus* (Gould et al. 2002) ABLV strains have been published. The genome size of the *Pteropus* ABLV strain of 11,918 nucleotides (nt) is the smallest lyssavirus for which complete genome information is available. All lyssaviruses sequenced to date have an even number of nucleotides. The significance of this observation will have to be established by sequencing of more lyssavirus genomes. The *Pteropus* strain has a 72.4% nucleotide sequence identity with RABV (SAD B19 strain), revealing the relatively close relationship between the two viruses. There are five putative reading frames corresponding consecutively to the nucleocapsid (N), phosphoprotein (P), matrix (M), glycoprotein (G) and polymerase (L) proteins of RABV. A striking feature of the genome structure is the long non-coding region between the glycoprotein and polymerase coding regions. It has previously been suggested that this is a remnant of an ancestral gene (Tordo et al. 1986). However, the absence of a downstream transcriptional termination signal in the region, as has been noted in some RABV strains, suggests a long 3′-noncoding region for the glycoprotein (Ravkov et al. 1995).

Transcription regulatory signals matched the consensus sequences of RABV (Conzelmann et al. 1990; Tordo et al. 1986) with the exception of single nucleotide variations occurring in the stop signal of the nucleocapsid protein of the pteropid isolate (Gould et al. 1998) and the stop signal for the phosphoprotein of the human isolate (Warrilow et al. 2002). Interestingly, in the latter case, sequencing of the same region of the original infected brain revealed the variation in the stop signal was absent. It was, therefore, speculated that a change from the consensus transcriptional stop signal may have been an adaptation to cell culture. Variation from the consensus for this signal has been noted in Mokola virus (MOKV) and shown to produce a bi-cistronic read-through product (Bourhy et al. 1990). How this may have provided an advantage to viral replication in culture is unknown.

The *Pteropus* strain has 70 and 131 nucleotide 3′ and 5′ untranslated regions (UTR), respectively, that correspond to the leader and trailer regions of rhabdoviruses (Warrilow et al. 2002). In a comparison with the UTRs of other rhabdoviruses, the 3′ UTR was evolutionarily more conserved than the 5′ UTR. The UTRs also indicated a closer relationship between ABLV and RABV than between ABLV and MOKV. A 9-nt sequence at the extreme 3′ and 5′ termini is complementary in ABLV, which is smaller than the 11-nt complementary termini of RABV and MOKV. ABLV termini complementarity extends to 23 of the 32 terminal nucleotides. The terminal 3′-UGC-5′, conserved in rhabdoviruses, is also present. The 3′UTR was U-rich, while the 5′ UTR was A-rich. These regions most likely contain important *cis*-acting regulatory signals for transcription and replication. The best-studied rhabdovirus in this regard is vesicular stomatitis virus (VSV). The reader is referred to a review for more detailed information (Barr et al. 2002).

At the amino acid level, all five putative viral proteins of the ABLV *Pteropus* and *Saccolaimus* strains have greater sequence homology to each other than to RABV. Pairwise comparison of the two ABLV strains reveals the order of evolutionary conservation of the proteins to be N >L >M >G >P, which is also reflected in the relationship between the viral proteins of ABLV and RABV. As no experimental data on ABLV proteins are available, elucidation of the function of viral proteins is limited to analogy with the other lyssaviruses, primarily RABV. The first open reading frame (ORF) codes for a putative polypeptide of 450 amino acids (51 K), the putative nucleocapsid protein. Alignment of the amino acid sequence of this ABLV protein with RABV (92%) revealed the close relationship between these viruses (Gould et al. 1998). The amino and carboxy terminals were the least conserved in this alignment. It has been suggested these may contain important B and T cell epitopes by analogy with RABV nucleocapsid protein. Both strains have a predicted phosphorylation site at Ser^{389} (Gould et al. 1998).

The second ORF encodes a 297 amino acid protein (33 K), the putative phosphoprotein. This is the least conserved of the viral proteins, with 83% amino acid homology between the ABLV strains (Gould et al. 2002). Four domains alternating in their degree of sequence variation were identified (Gould et al. 1998). The most conserved were between amino acids 1–52 and 174–254. The protein was most variable between amino acids 53–78 in an alignment with RABV, and amino acids 155–178 in an alignment between the two ABLV strains. A motif thought to be important in binding nucleocapsid protein, FSKKYKF, is present at amino acids 209–215. Another motif, $^{N}/_{S}$KXTQT, which resides at amino acids 143–148 and may be involved in binding the dynein light chain LC8, differs from isolates from genotypes 1, 4, 5 and 6 in the N-terminal reside of the motif (Nadin-Davis et al. 2002).

Two possible in-frame initiation codons are present in the third ORF, which encodes the putative matrix protein, for both strains of ABLV (Gould et al. 1998, 2002). The more downstream of these two codons is the likely authentic initiator, as the polyadenylation signal for the upstream phosphoprotein would overlap with the start of the matrix ORF, which is unlikely as this has not been observed for this group of negative-sense RNA viruses. Assuming an active downstream initiator, a matrix protein of 202 amino acids (23 K) would be translated. This protein has a high homology with RABV (87%) and between ABLV strains (92%) at the amino acid level. It has a conserved hydrophilic central region and a charged amino terminus. A series of four Asp residues from amino acids 27–30, which are conserved in lyssaviruses, are also present in ABLV.

The fourth ORF, the putative glycoprotein, is the only ABLV protein reported to vary in size between isolates of the same strain. Lyssavirus glycoprotein consists of a 19 amino acid signal peptide, a 439 amino acid ectodomain, a 22 amino acid transmembrane domain and an endodomain with varying sizes from 42–53 amino acids (Badrane et al. 2001). A putative N-glycosylation signal present at amino acid 319, which is conserved in all lyssaviruses, is also present. Due to probable post-translational processing including signal sequence cleavage and glycosylation, a predicted weight for the mature protein cannot be determined. However, the size of the complete glycoprotein ORF of *Pteropus* and *Saccolaimus* isolates are reported to be either 525 (Gould et al. 1998, 2002) or 526 (Guyatt et al. 2003; Warrilow et al. 2002) amino acids in length, respectively. This size difference is intriguing and could reflect the variable nature of the endodomain in the other lyssaviruses. *Pteropus* ABLV and RABV glycoprotein share 75% homology, while the two ABLV strains are 89% homologous. An analysis of synonymous and non-synonymous nucleotide substitutions indicated that most changes in the glycoprotein were purifying (Guyatt et al. 2003). Furthermore, evidence for positive selection was found for amino acids 499 and 501 that reside in the endodomain.

Thefifth ORF, the putative polymerase, is the largest of the viral proteins and encodes a protein of 2,127 amino acids (243 K). Six conserved blocks (I–VI) with sequences presumably important for polymerase function are present (Warrilow et al. 2002). The motif QGDNQ, analogous to the GDD motif of positive-strand RNA viruses, is present in block III. A leucine-zipper motif $LX_6LX_6LX_6L$ is reported to be present in the polymerases from amino acids 237–258 and 238–260 of the *Pteropus* and *Saccolaimus* strains, respectively. This motif is not present in the RABV polymerase but is present in MOKV. It will be of interest to determine if this motif has a role in polymerase function.

9
Taxonomy

Molecular evidence for ABLV being a member of the family *Rhabdoviridae*, genus *Lyssavirus* was initially provided by alignment of the deduced nucleocapsid protein sequence with other lyssaviruses (Fraser et al. 1996) and later by phylogenetic analysis of the same region (Gould et al. 1998, 2002). Further support was provided by phylogenetic analysis of the glycoprotein (Guyatt et al. 2003; Johnson et al. 2002), phosphoprotein (Nadin-Davis et al. 2002) and polymerase (Warrilow et al. 2002). All these analyses place both *Pteropus* and *Saccolaimus* ABLV strains in a separate group, genotype 7, when compared to the other six recognized genotypes in the genus (Fig. 2). On the basis of pathogenicity and immunogenicity, a further sub-division of the lyssaviruses into two phylogroups has been suggested. ABLV was included in phylogroup I, the division characterized by lyssaviruses that cross-neutralize each other and are pathogenic by the peripheral and intracerebral routes, as opposed to the intracerebral route only (Badrane et al. 2001).

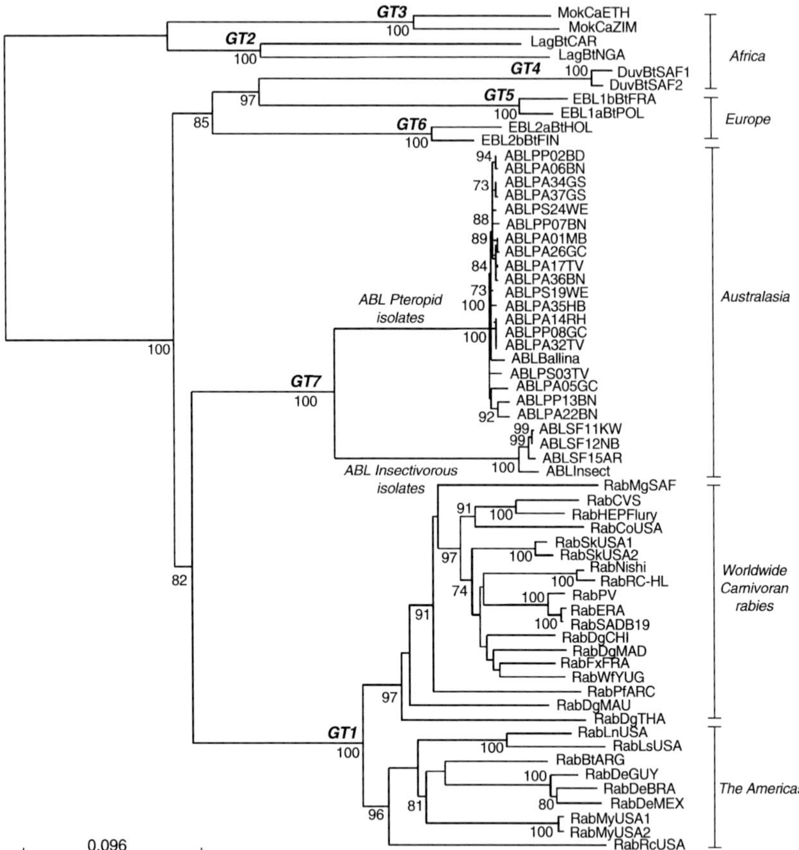

Fig. 2 Neighbour-joining phylogenetic tree demonstrating the genetic relationship of the genus *Lyssavirus* based on the G regions of the 61 lyssavirus isolates. *Numbers* at each node represent the percentage bootstrap support (1,000 replicates). Only bootstrap values ≥70% are indicated. Branch length is proportional to evolutionary distance between different lyssavirus isolates. The main phylogenetic groupings and their geographical locations are shown to the right of the figure (Reproduced with permission from Guyatt et al., Journal of General Virology 84, 485–496, 2003)

10 Host Biology and Distribution

Bats are classified into two sub-orders, *Megachiroptera* and *Microchiroptera*. There are thirteen species in the sub-order *Megachiroptera*, Family *Pteropodidae* living on continental Australia and its offshore islands (Hall and Richards

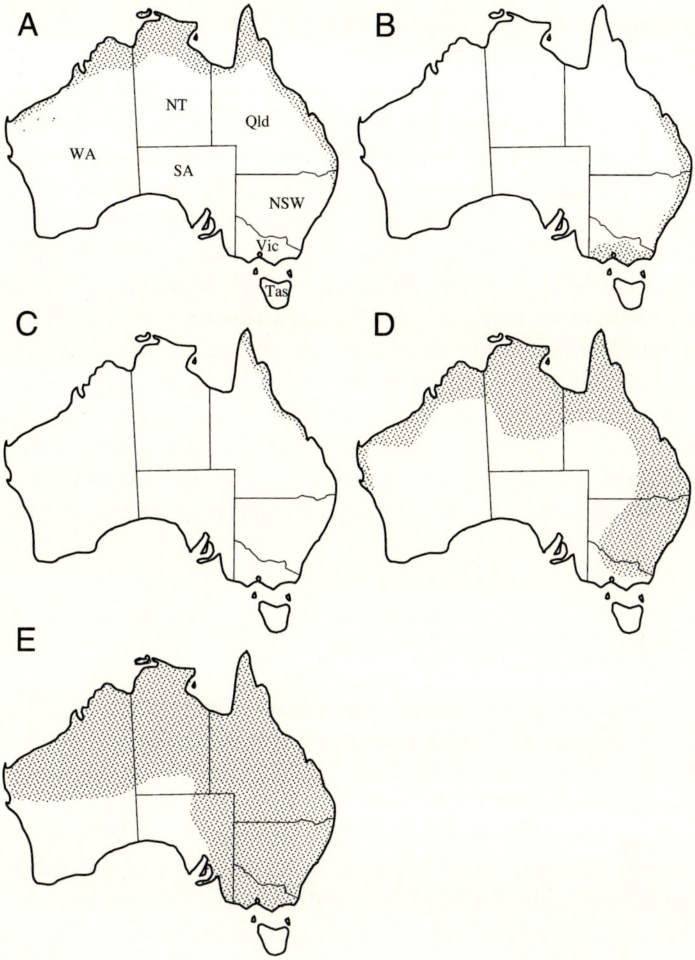

Fig. 3A–E Distribution of bats species that have been shown to be infected with Australian bat lyssavirus. A. *P. alecto*. B. *P. poliocephalus* C. *P. conspicillatus*. D. *P. scapulatus*. E. *S. flaviventris*. (Compiled from Churchill 1998). Australian states and territories shown are New South Wales (*NSW*), Northern Territory (*NT*), Queensland (*Qld*), South Australia (*SA*), Tasmania (*Tas*), Victoria (*Vic*) and Western Australia (*WA*)

2000). These animals are commonly referred to as fruit bats. Members of the genus *Pteropus* are commonly referred to as flying foxes. Four flying fox species: the black flying fox (*Pteropus alecto*), the grey-headed flying fox (*Pteropus poliocephalus*), the little red flying fox (*Pteropus scapulatus*) and the spectacled flying fox (*Pteropus conspicillatus*), have all been shown to be infected with the *Pteropus* strain of ABLV (McColl et al. 2000). The combined geographical distribution of these four species extends from Victoria, through New South Wales, Queensland, the Northern Territory and northwest Western Australia (Fig. 3). It is likely that the migration patterns of some Australian flying foxes may extend into Asia. Flying foxes live in camps consisting of possibly many thousands of individuals and sometimes more than one species (Churchill 1998). Consequently there exists an opportunity for cross-species transmission. There is circumstantial evidence that cross-species transmission can occur (Warrilow et al. 2003) and it has been suggested to explain the lack of genetic variation seen in the flying fox isolates.

The *Microchiroptera* are physically smaller and generally insectivorous animals that navigate by echolocation (Hall and Richards 2000), of which 65 species have been identified in Australia. One species, *Saccolaimus flaviventris*, has been shown to be infected with a different strain of ABLV to that of the *Pteropus* strain (Gould et al. 2002), and there is an unpublished observation of ABLV infection in other Microchiropteran species (Mackenzie et al. 2003).

11 Surveillance

Soon after the emergence of ABLV, Australian agricultural and health authorities initiated surveillance programs comprised of passive and active testing of wild bat populations for ABLV. Seroprevalence, indicating exposure but not necessarily active infection, was reported to be as high as 16% in a mixed group of 81 sick and healthy flying foxes (Hooper et al. 1997). In addition, some of the bats with neutralizing antibody to ABLV were apparently healthy.

From the available published information, ABLV-infected bats have been reported from the following states: Queensland, New South Wales, Victoria and the Northern Territory (Garner and Bunn 1997; Gould et al. 2002; Hooper et al. 1997; McCall et al. 2000; Tidemann et al. 1997; Warrilow et al. 2003). The only information on the prevalence of infected bats comes from passive surveillance of sick or injured bats, or bats involved in a scratch or bite incident with a human. One study of data tabulated nationally reported 7% of fruit bats and 1% of insectivorous bats, from a primarily sick sub-population, were infected (Garner and Bunn 1997). The great majority of bats tested were from three states: Queensland (80%), New South Wales (13%), Victoria (2.8%), and the Northern Territory (2.4%). In another two studies from Queensland, infected flying foxes were infected in 6% (McCall et al. 2000) or 9.4% (Warrilow et al. 2003) of submissions; no antigen positive *Microchiroptera* were reported in these studies. While no positive bats have been reported from

South Australia, Western Australia or Tasmania, all Australian bats should be considered possibly infected with ABLV. Due to bias in the bat populations in these passive surveillance studies, it is expected that the incidence of infection in wild populations would be much lower.

The detection of neutralizing antibodies in wild bats in the Philippines (Arguin et al. 2002) indicates that ABLV, or a closely related virus, is present in Asia, as was suggested previously (St George 1997). It was unfortunate that of 821 bats tested for lyssavirus infection by immunofluorescence assay of brain tissue none were positive, as it may have enabled an isolate to be obtained. Furthermore, there is an unpublished observation of a limited survey undertaken in Malaysia and Papua New Guinea in which no ABLV infection in bats could be demonstrated (Mackenzie et al. 2003). Further surveillance in the Asian region is required to determine the distribution of ABLV or closely related viruses.

12 Evolution

Assuming an origin external to Australia for the lyssaviruses, there are two possible scenarios for the entry of ABLV into Australia. The first scenario is based on the hypothesis that the common ancestor of the two strains resided within Australia. This is the most parsimonious scenario as it requires a single entry event. Applying this scenario enabled a minimum time estimate that the virus may have existed in Australia for at least 950 or 1,700 years, based on the rate of synonymous and non-synonymous nucleotide changes in the glycoprotein coding sequence, respectively (Warrilow et al. 2002).

Inconsistent with this first scenario is evidence suggestive of a more recent entry of ABLV into Australia. Australia's indigenous human population consume flying foxes as part of their traditional diet. Assuming that this is a practice that pre-dates European settlement, one might expect some traditional knowledge indicating that there is a risk of interacting with these animals. Tidemann (1997) makes the point that there is no such knowledge associated with hunting flying foxes. This is in contrast with traditional knowledge of risk of interacting with bats by indigenous peoples in Nigeria, the Pacific Northwest, Central and South America. Considering these traditional practices by indigenous Australians happened before our awareness of ABLV, and continue to happen, it is surprising that these practices have not resulted in a reported case of rabies-like encephalitis.

The number of scratch or bite incidents reported to public health authorities indicates contact between bats and humans is fairly common (McCall et al. 2000; Warrilow et al. 2003). There is no reason to think that it would have been otherwise prior to the first reported human case. Considering the frequency of infected bats involved in these incidents (6%–9.4%), if ABLV has been present in the bat populations for a long period, it is surprising that it has resulted in so few cases of rabies-like encephalitis. Another observation inconsistent with the first scenario is the lack of sequence variability in *Pteropus* isolates across Australia (Guyatt et al. 2003; Hooper et al. 1997), suggesting recent entry.

By an alternative scenario, the common ancestor of the two strains resided outside Australia. After divergence both lineages entered Australia independently at unknown times. By this scenario, the *Pteropus* strain may have entered Australia recently, possibly by migration of infected flying foxes from Asia. It is not possible to draw any conclusions from the *Saccolaimus* strain at present, due to the small number of isolates obtained, nearly all of which originated from southeast Queensland. Unfortunately, there is a lack of archival survey material that might provide evidence for the long-term existence of ABLV in Australia. Continued surveillance in Australia and Asia will hopefully increase our understanding of the distribution and evolution of the virus.

13 Public Health Issues and Conclusion

The similarity between ABLV and RABV, the ability of vaccines to afford cross-protection in mice, and the neutralizing properties of human rabies immune sera led to the adoption of rabies pre- and post-exposure protocols for the prevention of ABLV infection in humans. According to Australian government health guidelines, it is recommended that individuals at risk of exposure, such as veterinary workers and wildlife handlers, should undergo a course of three intramuscular injections with inactivated human diploid cell-cultured virus vaccine (MacIntyre 2000). Vaccinated individuals exposed to either bat scratch, bite or on mucous membranes should be offered a booster vaccination. Exposed unvaccinated individuals should be offered human rabies immune sera in addition to five intramuscular injections of vaccine. The exception to these post-exposure protocols is when the offending bat has been demonstrated to be uninfected within 48 h of the exposure, by a recognized laboratory. These protocols have most probably prevented further human cases.

In countries where classical rabies is a public health issue, the virus is usually present in domestic and wild carnivore populations, primarily dogs. Hence, there is the additional concern that ABLV may spill-over into carnivores and establish a new reservoir; however, the likelihood of this occurring seems remote. When spill-over events occur, they usually result in a dead-end infection in the new host. Australia's ecology has undergone dramatic change since European settlement. It remains to be seen to what extent these changes could assist the establishment of a permanent infectious cycle in a non-bat host following a possible spill-over event.

What more do we know about ABLV since the early work on its characterization, prevalence, vaccine coverage and pathology? The answer is, unfortunately, very little. Some small inroads have been made on understanding its molecular biology with sequence of both strains now available. However, we are no clearer in our understanding of how long ABLV has been in Australia. Experimental infection of bats may provide the means to determine how transmission in bats is maintained. Evidence suggesting that ABLV, or a closely related virus, is present in Southeast Asia, while not unexpected, has broadened our horizons. Surveillance results, particularly in flying foxes, indicate that the virus poses a risk to human health, and will probably continue to do so. One possible control measure that could reduce exposure to ABLV is vaccination of the reservoir chiropteran species to break the cycle of infection in the wild. The technical aspects of this approach have been discussed elsewhere (Mackenzie et al. 2003). However, considering the success of pre- and post exposure prophylaxis protocols, the enormous technical challenges involved in eliminating the virus from wild bat populations makes such an approach unfeasible with current technology. Passive surveillance efforts present the most effective means of monitoring bat populations and may reveal new affected bat species, new isolates, and hopefully a better understanding of the relationship between the virus and Australia's bats.

Acknowledgements The author would like to thank Kimberly Guyatt, Alan Gould and Alex Hyatt for contributing figures, and the staff at Queensland Health Scientific Services for assistance and support.

References

Allworth A, Murray K, Morgan J (1996) A human case of encephalitis due to a lyssavirus recently identified in fruit bats. Commun Dis Intell 20:504

Arguin PM, Murray-Lillibridge K, Miranda ME, Smith JS, Calaor AB, Rupprecht CE (2002) Serologic evidence of Lyssavirus infections among bats, the Philippines. Emerg Infect Dis 8:258–262

Badrane H, Bahloul C, Perrin P, Tordo N (2001) Evidence of two Lyssavirus phylogroups with distinct pathogenicity and immunogenicity. J Virol 75:3268–3276

Barr JN, Whelan SP, Wertz GW (2002) Transcriptional control of the RNA-dependent RNA polymerase of vesicular stomatitis virus. Biochim Biophys Acta 1577:337–353

Bisseru B (1972) Rabies. William Heinemann Medical Books Ltd, London, pp 18

Bourhy H, Sureau P, Tordo N (1990) From rabies to rabies-related viruses. Vet Microbiol 23:115–128

Churchill S (1998) Pteropodidae Australian Bats. New Holland Publishers, Sydney, pp 72–93

Conzelmann KK, Cox JH, Schneider LG, Thiel HJ (1990) Molecular cloning and complete nucleotide sequence of the attenuated rabies virus SAD B19. Virology 175:485–499

Faoagali JL, De Buse P, Strutton GM, Samaratunga H (1988) A case of rabies. Med J Aust 149:702–707

Field H, McCall B, Barrett J (1999) Australian bat lyssavirus infection in a captive juvenile black flying fox. Emerg Infect Dis 5:438–440

Fraser GC, Hooper PT, Lunt RA, Gould AR, Gleeson LJ, Hyatt AD, Russell GM, Kattenbelt JA (1996) Encephalitis caused by a Lyssavirus in fruit bats in Australia. Emerg Infect Dis 2:327–331

Garner G, Bunn C (1997) Update on surveillance for Australian bat lyssavirus. Aust Epidem 4:27–30

Gerrard J (1997) Fatal encephalitis and meningitis at the Gold Coast Hospital, 1980 to 1996. Commun Dis Intell 21:32–33

Gould AR, Hyatt AD, Lunt R, Kattenbelt JA, Hengstberger S, Blacksell SD (1998) Characterisation of a novel lyssavirus isolated from Pteropid bats in Australia. Virus Res 54:165–187

Gould A, Kattenbelt J, Gumley S, Lunt R (2002) Characterisation of an Australian bat lyssavirus variant isolated from an insectivorous bat. Virus Res 89:1–28

Grattan-Smith PJ, O'Regan WJ, Ellis PS, O'Flaherty SJ, McIntyre PB, Barnes CJ (1992) Rabies. A second Australian case, with a long incubation period. Med J Aust 156:651–654

Guyatt KJ, Twin J, Davis P, Holmes EC, Smith GA, Smith IA, Mackenzie JS, Young PL (2003) A molecular epidemiological study of Australian bat lyssavirus. J Gen Virol 84:485–496

Hall L, Richards G (2000) Introduction flying foxes: fruit and blossom bats of Australia. University of New South Wales Press, Sydney

Halpin K, Young PL, Field HE, Mackenzie JS (2000) Isolation of Hendra virus from pteropid bats: a natural reservoir of Hendra virus. J Gen Virol 81:1927–1932

Hanlon CA, DeMattos CA, DeMattos CC, Niezgoda M, Hooper DC, Koprowski H, Notkins A, Rupprecht CE (2001) Experimental utility of rabies virus-neutralizing human monoclonal antibodies in post-exposure prophylaxis. Vaccine 19:3834–3842

Hanna JN, Carney IK, Smith GA, Tannenberg AE, Deverill JE, Botha JA, Serafin IL, Harrower BJ, Fitzpatrick PF, Searle JW (2000) Australian bat lyssavirus infection: a second human case, with a long incubation period. Med J Aust 172:597–599

Heaton PR, Johnstone P, McElhinney LM, Roy C, O'Sullivan E, Whitby JE (1997) Heminested PCR assay for detection of six genotypes of rabies and rabies-related viruses. J Clin Microbiol 35:2762–2766

Hooper PT, Gould AR, Russell GM, Kattenbelt JA, Mitchell G (1996) The retrospective diagnosis of a second outbreak of equine morbillivirus infection. Aust Vet J 74:244–245

Hooper PT, Lunt RA, Gould AR, Samaratunga H, Hyatt AD, Gleeson LJ, Rodwell BJ, Rupprecht CE, Smith JE, Murray PK (1997) A new lyssavirus—the first endemic rabies related virus recognized in Australia. Bull Inst Pasteur 95

Hooper PT, Fraser GC, Foster RA, Storie GJ (1999) Histopathology and immunohistochemistry of bats infected by Australian bat lyssavirus. Aust Vet J 77:595–599

Johnson N, McElhinney LM, Smith J, Lowings P, Fooks AR (2002) Phylogenetic comparison of the genus Lyssavirus using distal coding sequences of the glycoprotein and nucleoprotein genes. Arch Virol 147:2111–2123

Jong K (1997) Lyssavirus infection study. New South Wales Public Health Bulletin 8:10

Lambert SB, Dalton CB, Carnie JA (1997) A review of fatal encephalitis in Victoria, 1993–1996. Australian Zoologist 4:20

MacIntyre CR (ed) (2000) The Australian Immunisation handbook, 7th edn. Australian Government Publishing Service, Canberra, pp 77–87

Mackenzie JS, Field HE, Guyatt KJ (2003) Managing emerging diseases borne by fruit bats (flying foxes), with particular reference to henipaviruses and Australian bat lyssavirus. J Appl Microbiol 94:59–69

McCall BJ, Epstein JH, Neill AS, Heel K, Field H, Barrett J, Smith GA, Selvey LA, Rodwell B, Lunt R (2000) Potential exposure to Australian bat lyssavirus, Queensland, 1996–1999. Emerg Infect Dis 6:259–264

McColl KA, Gould AR, Selleck PW, Hooper PT, Westbury HA, Smith JS (1993) Polymerase chain reaction and other laboratory techniques in the diagnosis of long incubation rabies in Australia. Aust Vet J 70:84–89

McColl KA, Tordo N, Aguilar Setien AA (2000) Bat lyssavirus infections. Rev Sci Tech 19:177–196

McColl KA, Chamberlain T, Lunt RA, Newberry KM, Middleton D, Westbury HA (2002) Pathogenesis studies with Australian bat lyssavirus in grey-headed flying foxes (*Pteropus poliocephalus*). Aust Vet J 80:636–641

Murray K, Rogers R, Selvey L, Selleck P, Hyatt A, Gould A, Gleeson L, Hooper P, Westbury H (1995a) A novel morbillivirus pneumonia of horses and its transmission to humans. Emerg Infect Dis 1:31–33

Murray K, Selleck P, Hooper P, Hyatt A, Gould A, Gleeson L, Westbury H, Hiley L, Selvey L, Rodwell B et al (1995b) A morbillivirus that caused fatal disease in horses and humans. Science 268:94–97

Nadin-Davis SA, Abdel-Malik M, Armstrong J, Wandeler AI (2002) Lyssavirus P gene characterisation provides insights into the phylogeny of the genus and identifies structural similarities and diversity within the encoded phosphoprotein. Virology 298:286–305

O'Sullivan JD, Allworth AM, Paterson DL, Snow TM, Boots R, Gleeson LJ, Gould AR, Hyatt AD, Bradfield J (1997) Fatal encephalitis due to novel paramyxovirus transmitted from horses. Lancet 349:93–95

Ravkov EV, Smith JS, Nichol ST (1995) Rabies virus glycoprotein gene contains a long 3′ noncoding region which lacks pseudogene properties. Virology 206:718–723

Rogers RJ, Douglas IC, Baldock FC, Glanville RJ, Seppanen KT, Gleeson LJ, Selleck PN, Dunn KJ (1996) Investigation of a second focus of equine morbillivirus infection in coastal Queensland. Aust Vet J 74:243–244

Samaratunga H, Searle JW, Hudson N (1998) Non-rabies Lyssavirus human encephalitis from fruit bats: Australian bat Lyssavirus (pteropid Lyssavirus) infection. Neuropathol Appl Neurobiol 24:331–335

Schneider LG, Dietzschold B, Dierks RE, Matthaeus W, Enzmann PJ, Strohmaier K (1973) Rabies group-specific ribonucleoprotein antigen and a test system for grouping and typing of rhabdoviruses. J Virol 11:748–755
Selvey LA, Wells RM, McCormack JG, Ansford AJ, Murray K, Rogers RJ, Lavercombe PS, Selleck P, Sheridan JW (1995) Infection of humans and horses by a newly described morbillivirus. Med J Aust 162:642–645
Skerratt LF, Speare R, Berger L, Winsor H (1998) Lyssaviral infection and lead poisoning in black flying foxes from Queensland. J Wildl Dis 34:355–361
Skull SA, Krause V, Dalton CB, Roberts LA (1999) A retrospective search for lyssavirus in humans in the Northern Territory. Aust N Z J Public Health 23:305–308
Speare R, Skerratt L, Foster R, Berger L, Hooper P, Lunt R, Blair D, Hansman D, Goulet M, Cooper S (1997) Australian bat lyssavirus infection in three fruit bats from north Queensland. Commun Dis Intell 21:117–120
St George TD (1997) Australian bat lyssavirus. Aust Vet J 75:367
Tidemann CR, Vardon MJ, Nelson JE, Speare R, Gleeson LJ (1997) Health and conservation implications of Australian bat Lyssavirus. Australian Zoologist 30:369–376
Tordo N, Poch O, Ermine A, Keith G, Rougeon F (1986) Walking along the rabies genome: is the large G-L intergenic region a remnant gene? Proc Natl Acad Sci U S A 83:3914–3918
Warrilow D, Smith IL, Harrower B, Smith GA (2002) Sequence analysis of an isolate from a fatal human infection of Australian bat lyssavirus. Virology 297:109–119
Warrilow D, Harrower B, Smith IL, Field H, Taylor R, Walker C, Smith GA (2003) Public health surveillance for Australian bat lyssavirus, in Queensland, Australia, 2000–2001. Emerg Infect Dis 9:262–264
Young PL, Halpin K, Selleck PW, Field H, Gravel JL, Kelly MA, Mackenzie JS (1996) Serologic evidence for the presence in *Pteropus* bats of a paramyxovirus related to equine morbillivirus. Emerg Infect Dis 2:239–240

CTMI (2005) 292:45–56

Pathogenesis of Rabies

B. Dietzschold[1] (✉) · M. Schnell[2] · H. Koprowski[1]

[1]Department of Microbiology, Center for Neurovirology,
Thomas Jefferson University, Philadelphia PA, USA
Bernhard.dietzschold@jefferson.edu

[2]Department of Biochemistry and Molecular Pharmacology,
Thomas Jefferson University, Philadelphia PA, USA

Abstract Rabies is a central nervous system (CNS) disease that is almost invariably fatal. The causative agent is rabies virus (RV), a negative-stranded RNA virus of the rhabdovirus family. RV pathogenesis, like that of other viruses, is a multigenic trait. Recent findings indicate that in addition to the RV G protein viral elements that regulate gene expression, especially expression of the L gene, are also likely to play a role in RV pathogenesis. In vivo, RV infects almost exclusively neurons, and neuroinvasiveness is the major defining characteristic of a classical RV infection. A key factor in the neuroinvasion of RV is transsynaptic neuronal spread. While the ability of RV to spread from the post-synaptic site to the pre-synaptic site is mediated by the RV G protein, the RV P protein might be an important determinant of retrograde transport of the virus within axons. Although the mechanism(s) by which an RV infection cause(s) a lethal neurological disease are still not well understood, the most significant factor underlying the lethal outcome of an RV infection appears to be the neuronal dysfunction due to drastically inhibited synthesis of proteins required in maintaining neuronal functions.

1 Introduction

Rabies is the tenth most common lethal infectious disease causing approximately 60,000 annual deaths worldwide [24]. While dogs represent the major reservoir of rabies virus [28] in most developing countries, the situation in the Americas is much more complex, since large reservoirs of rabies viruses exist in many wild animal species [38]. Outbreaks of rabies infection in these terrestrial mammals are found in broad geographic areas across the USA.

Rabies is a central nervous system (CNS) disease that is almost invariably fatal. The causative agent is rabies virus (RV), a negative-stranded RNA virus of the rhabdovirus family, which has a relatively simple modular genome organization and encodes a nucleoprotein (N), a phosphorylated protein (P), a matrix protein (M), a single external surface glycoprotein (G), and an RNA-dependent RNA polymerase (L). Fixed and street RV strains differ significantly from each other in their ability to invade the CNS from a peripheral site and cause a lethal encephalitis (Fig. 1). For example, the mortality in immunocompetent mice infected intramuscularly with street rabies strains (e.g., Dog-4, SHB-17) is 100–10,000 times higher than those infected with fixed rabies virus strains (e.g., CVS-B2c, CVS-N2c). Furthermore, while virus strains associated with canines are usually transmitted by severe bites that deeply inoculate large amounts of virus into subcutaneous and muscle tissue, bat-associated rabies viruses are likely delivered in comparatively negligible amounts. It has been suggested that an epizootic in the silver-haired bat popu-

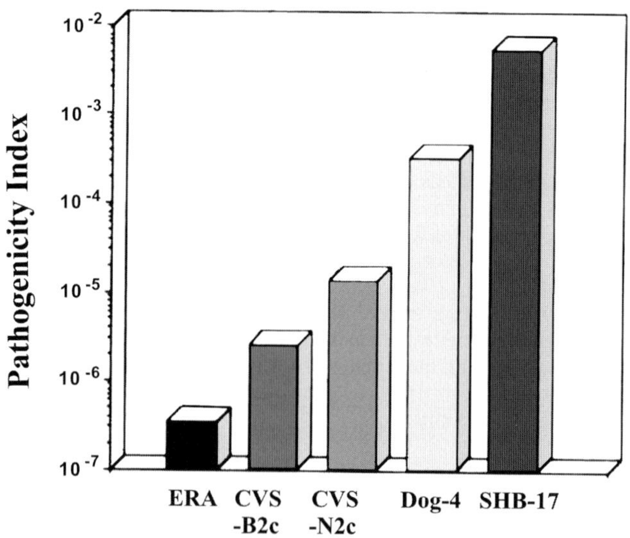

Fig. 1 Pathogenicity of tissue culture-adapted (*ERA*, *CVS-B2*) and mouse-adapted (*CVS-N2c*) fixed RV strains, and dog-associated (*Dog-4*) or silver-haired bat-associated (*SHB-17*) street RV strains. Pathogenicity index is defined as log intramuscular LD_{50} minus log virus titer determined in mouse neuroblastoma cells

lation might reflect adaptation of the virus to this species by either increasing its neuroinvasiveness or altering its tissue tropism, enabling transmission of disease by only a low number of virus particles [29]. Unfortunately, our current knowledge on RV pathogenesis is based only on either descriptive studies with street RV strains that provide limited insight into mechanisms underlying neurological disease, or on experiments using cell culture-adapted RV strains with questionable significance for the natural history of rabies. Nevertheless, the progress in cell biology and modern molecular biotechnology holds the promise of unraveling the mystery of the RV life cycle.

2 The Life Cycle of RV

Figure 2 illustrates the life cycle of RV. Although some cases of human rabies infection initiate at the respiratory tract, perhaps via aerosol transmission [6], the site of RV entry is usually the skin or mucosal membrane where the virus is introduced into the deeper layers of the skin or into muscle tissue through biting, licking, or scratching by an RV-infected mammal (usually a carnivore or bat) [4]. In vivo, RV infects neurons almost exclusively and the infection is cell-associated during nearly the entire life cycle with a few exceptions: (1) RV may replicate at the entry site in non-neuronal cells (e.g., muscle cells) [4] before it enters unmyelinated terminals of motor or sensory nerves; and (2) at the final stage of disease, RV infects acinar cells of the salivary glands, which produce and release large amounts of infectious virus particles together with the saliva [4] (Fig. 2). Since neurotropism, neuroinvasiveness, and impairment of neuronal functions are the three major defining characteristics of a classical RV infection [41], we discuss here molecular mechanisms that might play a role in these three RV hallmarks.

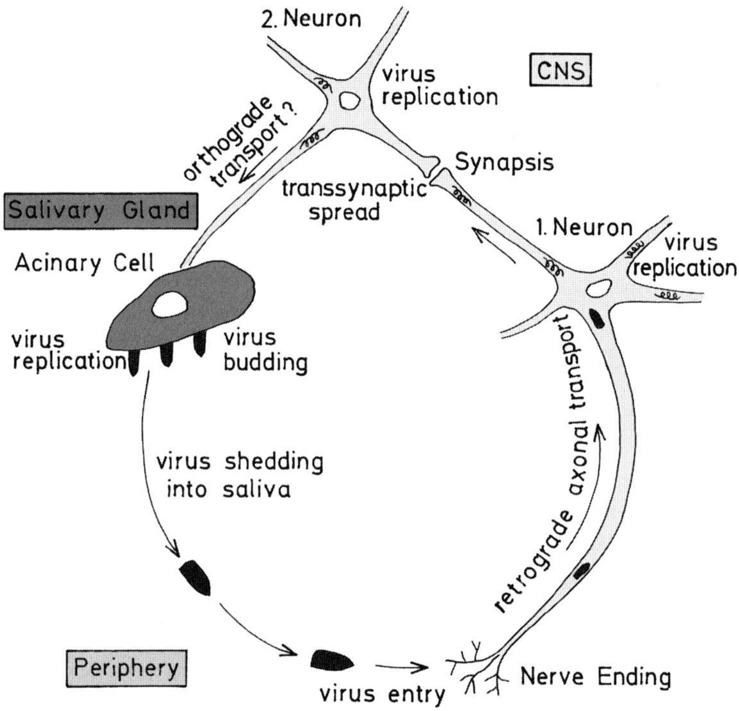

Fig. 2 The life cycle of rabies virus. RV infection starts at a peripheral site (e.g., skin or muscle tissue). The virus enters unmyelinated nerve terminals and migrates by retrograde axonal transport to the neuronal cell body. After replication in the cell body of the primary neuron, infection proceeds via retrograde axonal transport and transsynaptic spread through several neurons before it infects acinar cells which shed the virus into the oral cavity

3
Molecular Mechanisms That Determine the Neurotropism of RV

The interaction between the RV G and the putative host cell receptor is considered to be important for viral pathogenesis and so recent research on RV neurotropism has focused on the identification of viral receptors. Several observations have led to the hypothesis that the nicotinic acetylcholine receptor (nAChR) serves as a receptor for RV [14, 16, 21, 22]: (1) RV ribonucleoprotein (RNP) is detected at neuromuscular junctions; (2) amino acid residues 189–214 of RV G and snake venom neurotoxin are highly homologous; and (3) α-bungarotoxin or D-tubocurarine can block the attachment of RV to myotubes. However, RV can also infect neurons in vivo that do not express nAChRs. In this context, it has been reported that other molecules including phospholipids, gangliosides, neuronal cell adhesion molecule and the nerve growth factor receptor can act as RV receptors [8, 20, 37, 44, 45]. Thus, as with many other neurotropic viruses, the identity of an RV receptor remains controversial.

4 Putative Mechanisms of RV Neuroinvasiveness

Neuroinvasiveness refers to the ability of the virus to invade the nervous system from a peripheral site. The G protein of RV is a major contributor to this ability of the virus [7, 25, 30, 31, 40], and several G-associated pathogenic mechanisms have been identified: (1) G must interact effectively with cell surface molecules that can mediate rapid virus uptake [8]; (2) G must interact optimally with the RNA-NP-M complex for efficient virus budding [26, 31]; and (3) expression levels of G must be kept at a minimum to prevent functional impairment of the infected neuron [30]. The pathogenicity of fixed rabies virus strains (i.e., ERA, HEP, CVS) has been correlated with the presence of a determinant located in antigenic site III of the G protein [7, 40]. It has also been shown that mutation of Arg to Ileu, Gln, Glu, or Gly at position 333 of the G protein amino acid sequence completely abolishes virulence in adult immunocompetent mice regardless of the virus dose or route of inoculation [7, 40, 43]. Whereas the markedly reduced spread of these nonpathogenic mutant viruses within the CNS indicated the absolute requirement for Arg at position 333 of G in the spread of lethal RV in adult animals [7], the significance of this finding for the natural history of rabies is not clear since the mutants were obtained from highly attenuated tissue culture-adapted strains (e.g., ERA, CVS-11), which showed pathogenicity indices far lower than that of most street RV strains (e.g., dog- and bat-associated RV strains) (Fig. 1) despite an Arg_{333} in the G protein [30]. Furthermore, phenotypic analyses of recombinant RVs in which the G gene of a non-neuroinvasive and less neurotropic strain was replaced with that from highly neuroinvasive and neurotropic strains revealed in every case markedly lower pathogenicity of the recombinant viruses than that of the wild-type viruses [31]. This finding suggests that RV pathogenesis, like that of influenza virus [2, 5, 17] is a multigenic trait. Regulation of gene expression, especially expression of the L gene, is also likely to play a role in RV pathogenesis [11]. Besides the gene order, which is a major factor in transcriptional regulation of nonsegmented negative-strand RNA viruses [1, 12], specific gene border sequences that contribute considerably to the regulation of RV gene expression have been implicated in viral pathogenesis [11].

Another key factor in the neuroinvasion of RV is transsynaptic spread, which is the ability of the virus to use synaptic junctions to propagate within the CNS (Fig. 3). Whereas the immunoelectron microscopic detection of RV RNP but not RV G clusters at the postsynaptic site and in axons associated with neurofilaments suggested that RV G might not be necessary for transsynaptic spread [15], the reduced spread of RV antigenic site III mutants within the CNS demonstrated the requirement for a functionally intact G protein in axonal/transsynaptic spread of lethal RV infection [8]. In addition, Etessami

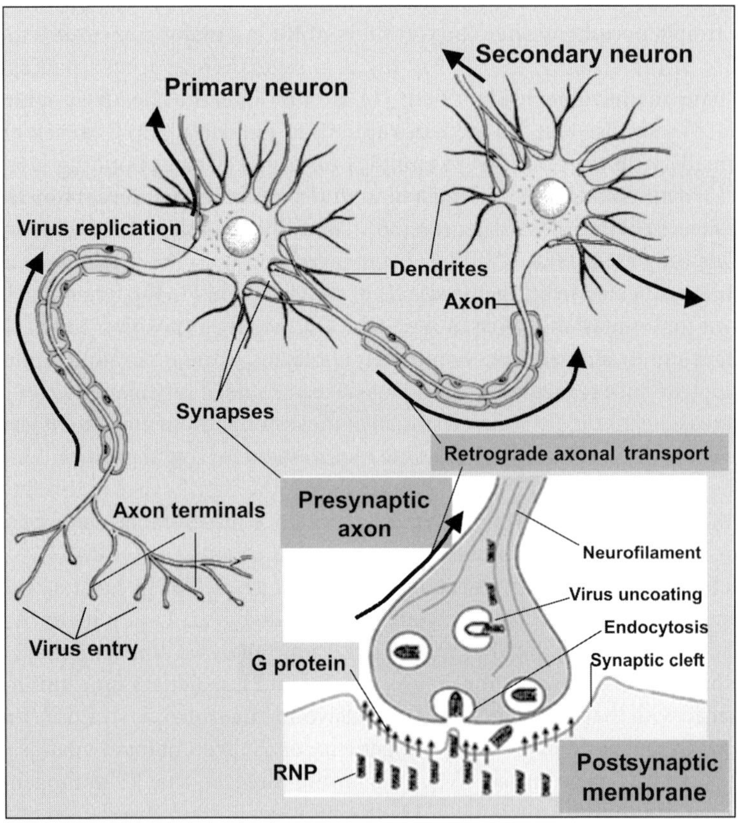

Fig. 3 Transsynaptic spread and retrograde axonal trafficking of RV. RV exits at the postsynaptic site and enters the presynaptic part probably via a receptor-mediated endocytosis. Following uptake, the RV particle-containing endosomes are fused, resulting in the liberation of RNP complexes which utilize dynein for retrograde transport. Alternatively, retrograde transport of virus-containing endosomes might occur before uncoating of the virus in the neuronal cell body

et al. [9] show that a G-deleted RV (ΔG) *trans*-complemented with RV G can infect neuronal cells but cannot spread to secondary neurons. While these experiments indicate the major role of the G protein in RV spread from the postsynaptic site to the presynaptic site, the RV P protein might be an important determinant of retrograde transport of the virus within axons. This hypothesis is supported by the demonstration that RV P interacts strongly with the dynein light chain LC8 [19, 35] via a conserved (K/R)*X*TQT motif [23]. Indeed, deletion of the LC8 binding region from RV P of the vaccine strain SAD-D29 led to a 30-fold decrease in the LD_{50} in 2-day-old suckling mice. However, the ΔLC8 RV still killed suckling mice at doses higher than 10^2 focus-forming units, suggesting that the LC8 binding site may not be the sole factor in the retrograde axoplasmic flow of RV [27]. Another unresolved issue concerning the potential RV P-mediated retrograde transport mechanism rests in the requirement for virus uncoating immediately after virus uptake at the axon terminal or at the presynaptic site (Fig. 3); to date, there is no clear evidence that virus uncoating actually takes place at these sites.

An alternative mechanism for the retrograde transport of RV might involve the trafficking of whole virus particles in the form of virus-containing vesicles. Such a mechanism is supported by the findings that the RV G can specifically bind to the neurotrophin receptor p75 ^{NTR} [46], and that ligand p75 ^{NTR} complexes are internalized via clathrin-coated pits into early endosomes, which then can traffic in a retrograde manner [3]. Thus, it is likely that RV-p75 ^{NTR} complexes might well be transported in a similar fashion.

5 Mechanisms Involved in RV-Induced Neuronal Damage

The ability of RV to produce neurological disease is not well understood, and the underlying mechanisms have eluded investigators for more than a century. It has long been known that human rabies patients show limited morphological brain damage, with no or only minimal neuronal cell death [32]. A mechanism contributing to the CNS dysfunction could be the impairment of neuronal functions, in particular, by the alteration of neurotransmission [42]. In this context, it has been shown that expression of housekeeping genes is markedly decreased in RV-infected neurons, resulting in a generalized inhibition of protein synthesis and, in turn, impairment of vital neuronal functions [13]. It has also been speculated that the expression of RV G protein on the cell surface might lead to a disturbance of ion channels, a hypothesis consistent with the electrophysiological alterations observed in the brains of RV-infected mice [42].

The induction of apoptosis in RV-infected neurons has recently been proposed as a potential pathogenic mechanism [18], although the use of an attenuated strain in studies on which this proposal was based raise questions about the relevance to street RV infection. Moreover, since apoptosis is likely a host defense mechanism that leads to a self-limiting virus infection, it is not clear why a rabies strain that is well-adapted to its natural host would trigger a mechanism detrimental to its own survival. In fact, adenoviruses [33] and human hepatitis C virus [36] contain genes that act to interfere with apoptosis, thus allowing cell survival and virus replication. Those findings and our own demonstration that the pathogenicity of a particular RV strain correlates inversely with its ability to induce programmed cell death [30] argue against any strict correlation between viral pathogenesis and virus-induced cell death. Together, the most significant factor underlying the lethal outcome of an RV infection appears to be the neuronal dysfunction due to drastically inhibited synthesis of proteins required in maintaining neuronal functions.

6 Reverse Genetics Analyses of Pathogenesis

The recent advent of reverse genetics technology [39] holds the promise of a better understanding of RV pathogenesis. The modular genomic organization of the virus readily allows genetic manipulations of viral genes and stable expression of large foreign genes up to 6 kb in size. For example, in experiments to determine whether the G protein regulates pathogenicity [31], the respective rabies virus G genes of the neuroinvasive and highly neurotropic strains SHBRV-18, CVS-N2c, and CVS-B2c were introduced into the non-neuroinvasive and less neurotropic SN-10 strain; phenotypic analyses of the recombinant viruses indicated that the neurotropism of a particular rabies virus strain was a function of its G. Nevertheless, the pathogenicity of the recombinant viruses was consistently much lower than that of the wild-type viruses, suggesting that G protein regulation of neurotropism is not the only attribute important in pathogenesis.

Reverse genetics technology has also been used to address the important question of whether induction of apoptosis plays a role in RV pathogenesis. In a study raising a recombinant virus constructed to express the pro-apoptotic protein cytochrome c (Cyto c), apoptosis in primary neuron cultures was markedly increased as compared with cultures infected with recombinant not expressing Cyto c [34]. Mortality in mice infected with the Cyto c-expressing construct was substantially lower than in mice infected with the Cyto c-nonexpressing construct, confirming the hypothesis that RV-induced cell death contributes to a decrease in pathogenicity [10, 34].

To investigate whether the ability of particular RV strains to induce apoptosis depends on the expression level of its G protein, a recombinant RV carrying two identical G genes (SPBNGA-GA) was constructed and used to infect tissue culture cells [10]; immunoprecipitation analysis and flow cytometry revealed on average twice the level of RV G as in cells infected with RV carrying only a single RV G gene (SPBNGA). Moreover, the overexpression of RV G was paralleled by a significant increase in caspase-3 activity followed by a marked decrease in mitochondrial respiration, neither of which was observed in SPBNGA-infected cells. Furthermore, fluorescence staining and confocal microscopy revealed an increase in apoptosis and markedly reduced neurofilament and F actin levels in SPBNGA-GA-infected primary neuron cultures as compared with the SPBNGA-infected cultures. These data support the concept that the ability of RV to trigger the apoptotic cascade depends on the G protein level. Clearly, the ability to genetically modify RV provides a powerful avenue to obtaining insight into RV pathogenesis.

7
Future Perspectives

Although recent advances in molecular biology have led to some insight into the mechanisms by which rabies virus infection causes neurological disease, the conceptual understanding RV pathogenesis remains incomplete and many of the molecular mechanisms involved are still unclear. In particular, further research on the mechanisms of RV retrograde transport and transsynaptic spread is necessary. Other open questions concern behavioral changes of the infected animal (e.g., aggressiveness) at the onset of the disease, which manifest an essential part of the RV life cycle, and the mode of RV trafficking to the salivary gland and shedding into the saliva before the animal becomes moribund. To better understand RV pathogenesis, future research efforts must focus on virus–host interactions, especially: (1) on cellular proteins that facilitate virus uptake by neurons and virus spread within the neuronal network; (2) on the role of RV replication and transcription (e.g., L protein, transcription signal sequences, and intergenic regions) in the pathogenesis; and (3) on the identification of signal transduction elements that are activated during RV infection (e.g., expression of RV proteins, such as G) and possibly result in calcium signaling to trigger a distal event with global consequences.

Acknowledgements This work was supported by Public Health Service Grants AI45097-6 and AI09706-32.

References

1. Ball LA, Pringle CR, Flanagan B, Perepelitsa VP, Wertz GW (1999) Phenotypic consequences of rearranging the P, M, G genes of vesicular stomatitis virus. J Virol 73:4705–4712
2. Basler CF, Reid AH, Dybing JK, Janczewski TA, Fanning TG, Zheng H, Salvatore M, Perdue ML, Swayne DE, Garcia-Sastre A, Palese P, Taubenberger JK (2001) Sequence of the 1918 pandemic influenza virus nonstructural gene (NS) segment and characterization of recombinant viruses bearing the 1918 NS genes. Proc Natl Acad Sci U S A 98:2746–2751
3. Butowt R, von Bartheld CS (2003) Connecting the dots: trafficking of neurotrophins, lectins and diverse pathogens by binding to the neurotrophin receptor p75NTR Eur J Neurosci 17:673–680
4. Charlton KM (1994) The pathogenesis of rabies and other lyssaviral infections: recent studies. In: Rupprecht CE, Dietzschold B, Koprowski H (eds) Lyssaviruses. Springer-Verlag, Berlin Heidelberg New York, pp 95–115
5. Chen W, Calvo PA, Malide D, Gibbs J, Schubert U, Bacik I, Basta S, O'Neill R, Schickli J, Palese P, Henklein P, Bennink JR, Yewdell JW (2001) A novel influenza A virus mitochondrial protein that induces cell death. Nature Med 7:1306–1332
6. Constatine DG (1967) Rabies transmission by air in bat caverns. CDC, Atlanta, Public Health Service publication 1617
7. Dietzschold B, Wunner WH, Wiktor TJ, Lopes AD, Lafon M, Smith CL, Koprowski H (1983) Characterization of an antigenic determinant of the glycoprotein that correlates with pathogenicity of rabies virus. Proc Natl Acad Sci U S A 80:70–74
8. Dietzschold B, Wiktor TJ, Trojanowski JQ, Macfarlan RI, Wunner WH, Torres-Anjel MJ, Koprowski H (1985) Differences in cell-to-cell spread of pathogenic and apathogenic rabies virus in vivo and in vitro. J Virol 56:12–18
9. Etessami R, Conzelmann K-K, Fadai-Ghotbi B, Natelson B, Tsiang H, Ceccaldi P-E (2000) Spread and pathogenic characteristics of a G-deficient rabies virus recombinant: an *in vitro* and *in vivo* study. J Gen Virol 81:2147–2153
10. Faber M, Pulmanausahakul R, Hodawadekar SS, Spitsin S, McGettigan JP, Schnell MJ, Dietzschold B (2002) Overexpression of the rabies virus glycoprotein (G) results in enhancement of apoptosis and anti-viral immune response. J Virol 76:3374–3381
11. Finke S, Cox JH, Conzelmann KK (2000) Differential transcription attenuation of rabies virus genes by intergenic regions: generation of recombinant viruses overexpressing the polymerase gene. J Virol 74:7261–7269
12. Flanagan EB, Ball LA, Wertz GW (2000) Moving the glycoprotein gene of vesicular stomatitis virus to promoter-proximal positions accelerates and enhances the protective immune response. J Virol 74:7895–7902
13. Fu ZF, Weihe E, Zheng YM, Schaefer MK-H, Sheng H, Corisdeo S, Rauscher FJ III, Koprowski H, Dietzschold B (1993) Differential effects of rabies and Borna disease viruses on immediate-early- and late-response gene expression in brain tissue. J Virol 67:6674–6681

14. Gastka M, Horvath J, Lentz TL (1996) Rabies virus binding to the nicotinic acetylcholine receptor α subunit demonstrated by the virus overlay protein binding assay. J Gen Virol 77:2437–2440
15. Gostonyi G (1994) Reproduction of lyssaviruses: ultrastructural composition of lyssavirus and functional aspects of pathogenesis. In: Rupprecht CE, Dietzschold B, Koprowski H (eds) Lyssaviruses. Springer-Verlag, Berlin Heidelberg New York, pp 43–68
16. Hanham CA, Zhao F, Tignor GH (1993) Evidence from the anti-idiotypic network that the acetylcholine receptor is a rabies virus receptor. J Virol 67:530–542
17. Hatta M, Gao P, Halfmann P, Kawaoka Y (2001) Molecular basis for high virulence of Hong Kong H5N1 influenza A viruses. Science 293:1840–1842
18. Jackson AC, Rossiter JP (1997) Apoptosis plays an important role in experimental rabies virus infection. J Virol 71:5603–5607
19. Jacob Y, Badrane H, Ceccaldi P-E, Tordo N (2000) Cytoplasmic dynein LC8 interacts with lyssavirus phosphoprotein. J Virol 74:10217–10222
20. Kucera P, Dolovo M, Coulon P, Flamand A (1985) Pathways of the early propagation of virulent and avirulent rabies strains from the eye to the brain. J Virol 55:158–162
21. Lentz TL, Burrage TG, Smith AL, Crick J, Tignor GH (1982) Is the acetylcholine receptor a rabies virus receptor? Science 215:182–184
22. Lentz TL, Wilson PT, Speicher DM (1984) Amino acid sequence similarity between rabies virus glycoprotein and snake venom curaremimetic neurotoxin. Science 226:847–848
23. Lo KW-H, Naisbitt S, Fan J-S, Sheng M Zhang M (2001) The 8-kDa dynein light chain binds to its targets via a conserved (K/RXTQT) motif. J Biol Chem 276:14059–14066
24. Martinez L (2000) Global infectious disease surveillance. Int J Infect Dis 4:222–228
25. Mebatsion T, Finke S, Weiland F, Conzelmann K-K (1997) A CXCR4/CD4 pseudotype rhabdovirus that selectively infects HIV-1 envelope protein-expressing cells. Cell 90:841–847
26. Mebatsion T, Weiland F, Conzelmann KK (1999) Matrix protein of rabies virus is responsible for the assembly and budding of bullet-shaped particles and interacts with the transmembrane spike glycoprotein G. J Virol 73:242–250
27. Mebatsion T (2001) Extensive attenuation of rabies virus by simultaneously modifying the dynein light chain binding site in the P protein and replacing Arg333 in the G protein. J Virol 75:11496–11502
28. Meslin F-X, Fishbein DB, Matter HC (1994) Rationale and prospects for rabies elimination in developing countries, In: Rupprecht CE, Dietzschold B, Koprowski H (eds) Lyssaviruses. Springer-Verlag, Berlin Heidelberg New York, pp 1–26
29. Morimoto K, Patel M, Corisdeo S, Hooper DC, Fu ZF, Rupprecht CE, Koprowski H, Dietzschold B (1996) Characterization of a unique variant of bat rabies virus responsible for newly emerging human cases in North America. Proc Natl Acad.Sci U S A 93:5653–5658
30. Morimoto K, Hooper DC, Spitsin S, Koprowski H, Dietzschold B (1999) Pathogenicity of different rabies virus variants inversely correlates with apoptosis and rabies virus glycoprotein expression in infected primary neuron cultures. J Virol 73:510–518
31. Morimoto K, Foley HD, McGettigan JP, Schnell MJ, Dietzschold B (2000) Reinvestigation of the role of the rabies virus glycoprotein in viral pathogenesis using a reverse genetics approach. J Neuro Virol 6:373–381
32. Murphy FA (1977) Rabies pathogenesis. Arch Virol 54:279–297

33. Pilder S, Logan J, Shenk T (1984) Deletion of the gene encoding the adenovirus 5 early region 1b 21,000-molecular-weight polypeptide leads to degradation of viral and host cell DNA. J Virol 52:664–671
34. Pulmanausahakul R, Faber M, Morimoto K, Spitsin S, Weihe E, Hooper DC, Schnell MJ, Dietzschold B (2001) Overexpression of cytochrome c by a recombinant rabies virus attenuates pathogenicity and enhances antiviral immunity. J Virol 75:10800–10807
35. Raux H, Flamand A, Blondel D (2000) Interaction of the rabies virus P protein with the LC8 dynein light chain. J Virol 74:10212–10216
36. Ray RB, Meyer K, Ray R (1996) Suppression of apoptotic cell death by hepatitis C virus core protein. Virology 226:176–182
37. Rupprecht CE, Dietzschold B, Koprowski H (1994) Lyssavirus. In: Current topics in microbiology and immunology. 187:1–339
38. Rupprecht CE, Smith JS, Fekadu M, Childs JE (1995) The ascension of wildlife rabies: a cause for public health concern or intervention? Emerg Infect Dis 1:107–114
39. Schnell MJ, Mebatsion T, Conzelmann KK (1994) Infectious rabies viruses from cloned cDNA. EMBO J 13:4195–4203
40. Seif I, Coulon P, Rollin PE, Flamand A (1985) Rabies virulence: effect on pathogenicity and sequence characterization of rabies virus mutations affecting antigenic site III of the glycoprotein. J Virol 53:926–934
41. Stroop WG (1995) Viral pathogenesis. In: McKendall RR, Stroop WG (eds) Handbook of Neurovirology. Marcel Deckker, New York, pp 27–54
42. Tsiang H (1982) Neuronal function impairment in rabies-infected rat brain. J Gen Virol 61:277–281
43. Tuffereau C, Leblois H, Benejean J, Coulon P, Lafay F, Flamand A (1989) Arginine or lysine in position 333 of ERA and CVS glycoprotein is necessary for rabies virulence in adult mice. Virology 172:206–212
44. Tuffereau C, Benejean J, Blondel D, Kieffer B, Flamand A (1998) Low-affinity nerve-growth factor receptor (P75NTR) can serve as a receptor for rabies virus. EMBO J 17:7250–7259
45. Tuffereau C, Benejean J, Alfonso A-MR, Flamand A, Fishman MC (1998) Neuronal cell surface molecules mediate specific binding to rabies virus glycoprotein expressed by a recombinant baculovirus on the surfaces of lepidopteran cells. J Virol 72:1085–1091
46. Tuffereau C, Desmezieres E, Benejean J, Jallet C, Flamand A, Tordo N, Perrin P (2001) Interaction of lyssaviruses with low-affinity nerve-growth factor receptor $p75^{NTR}$. J Gen Virol 82:2861–2867

CTMI (2005) 292:57–80

Bovine Ephemeral Fever in Australia and the World

P. J. Walker

CSIRO Livestock Industries, Australian Animal Health Laboratory (AAHL),
5 Portarlington Road, 3220 Geelong, Victoria, Australia
Peter.Walker@csiro.au

Abstract Bovine ephemeral fever (BEF) is a disabling viral disease of cattle and water buffaloes. It can cause significant economic impact through reduced milk production in dairy herds, loss of condition in beef cattle and loss of draught animals at the time of harvest. Available evidence indicates clinical signs of BEF, which include bi-phasic fever, anorexia, muscle stiffness, ocular and nasal discharge, ruminal stasis and recumbency, are due primarily to a vascular inflammatory response. In Australia, between 1936 and 1976, BEF occurred in sweeping epizootics that commenced in the tropical far north and spread over vast cattle grazing areas of the continent. In the late 1970s, following several epizootics in rapid succession, the disease became enzootic in most of northern and eastern Australia. In Africa, the Middle East and Asia, BEF occurs as also epizootics which originate in enzootic tropical areas and sweep north or south to sub-tropical and temperate zones. The causative virus is transmitted by haematophagous insects that appear to be borne on the wind, allowing rapid spread of the disease. *Bovine ephemeral fever virus* (BEFV) has been classified as the type species of the genus *Ephemerovirus* in the *Rhabdoviridae*. It has a complex genome organization which includes two glycoprotein genes that appear to have arisen by gene duplication. The virion surface glycoprotein (G protein) contains four major antigenic sites that are targets for neutralizing antibody. An analysis of a large number of BEFV isolates collected in Australia between 1956 and 1992 has indicated remarkable stability in most neutralization sites. However, epitope shifts have occurred in the major conformational site G3 and these have been traced to specific mutations in the amino acid sequence. BEFV isolates from mainland China and Taiwan are closely related to Australian isolates, but some variations have been detected. Natural BEFV infection induces a strong neutralizing antibody response and infection usually induces durable immunity. Several forms of live-attenuated, inactivated and recombinant vaccines have been reported but with variable efficacy and durability of protection. The BEFV G protein is a highly effective vaccine antigen, either as a purified subunit or expressed from recombinant viral vectors.

1 Introduction

Bovine ephemeral fever (BEF, also variously called three-day sickness, bovine enzootic fever, bovine influenza and stiffsiekte) is a disabling viral disease of cattle and water buffaloes that occurs seasonally in tropical, subtropical and temperate regions of Africa, Asia and Australia. The disease is often characterized by the rapid onset of and recovery from clinical signs that can include a bi-phasic fever, anorexia, muscle stiffness, ocular and nasal discharge, salivation, depression, rumenal stasis, lameness and sternal recumbency (St George 1990). Morbidity rates can vary from 1% to almost 100%, probably due to prevailing epidemiological factors rather than variations in viral virulence. More severe infections occur in heavy or lactating animals but mortality rates rarely exceed 1%. At the first sign of disease in dairy cattle, there is a sudden drop in milk production and this may not recover until the next lactation (Theodoridis et al. 1973b; Davis et al. 1984). There is also some evidence that BEF can cause temporary infertility in bulls and delayed oestrus and mid-term abortions in cows (Chenoweth and Burgess 1972; Theodoridis et al. 1973b; Uren et al. 1987). Loss in milk production, loss of condition in beef herds, infertility and temporary disablement of draught animals at the time of harvest cause significant economic loss to individual farmers. In large epizootics in Australia, the industry-wide economic impact of BEF has been estimated to exceed A$100 million (St George 1986). Evidence of ephemeral fever infection can also impact on trade in live animals and semen to disease-free zones (Uren 1989).

Aspects of bovine ephemeral fever, focussing particularly on diagnosis, pathology, treatment and control, have recently been reviewed (Nandi and Negi 1999; Kirkland 2002). This review will examine the history and epidemiology of BEF in Australia and other parts of the world. The molecular and antigenic structure of the virus will also be reviewed and the prospects for development of new, improved vaccines based on recombinant antigens will be discussed.

2 Origins of the Disease

According to St George (1988), the first reference to bovine ephemeral fever was a brief report of the occurrence of the disease during 1868–1871 in the German East Africa—a territory now occupied by Tanzania, Burundi and Rwanda (Schweinfurth 1878). However, the first detailed account was of a BEF epizootic in Egypt in 1895 described as "dengue of cattle" (Piot 1896) but with unmistakable characteristics of the modern disease. A major epizootic in Zimbabwe in 1907 led to the first experimental investigations including demonstration of the transmission of disease in infected cattle blood (Bevan 1907, 1912). In subsequent years to 1935, there were reports of BEF from Palestine, India, Indonesia and Japan (van der Westhuizen 1967). However, these reports were not associated with contemporaneous trans-global movements of ruminants or vectors. Indeed, ephemeral fever is believed to be an ancient disease that has been endemic in much of Africa and southern Asia since antiquity (St George 1988). The apparent emergence and re-emergence of the disease during the past 125 years is likely due to the development and growth of the cattle industry and improvements in veterinary services in Australia, Africa and Asia (St George and Standfast 1988).

3 History and Epizootiology of BEF in Australia

Bovine ephemeral fever appears to have first entered Australia, probably from the Indonesian Archipelago, more than 60 years ago. The first recorded disease outbreak occurred in February 1936 in the Humbert River district of the Northern Territory. During the following year, an epizootic swept westward into the Kimberly region of Western Australia, and eastward through a vast area of Queensland and New South Wales, reaching northern Victoria in the late summer of 1937. In northern regions, the prevalence of infection approached 100%. By May 1937, the epizootic had largely subsided (Seddon 1938). For almost 20 years following the epizootic, BEF occurred only as sporadic, localized outbreaks in far northern Australia.

In 1955–1956, a second sweeping BEF epizootic occurred. Disease was first reported in October 1955 at Millungera Station on the Flinders River in north-western Queensland. At about the same time, there were reports of disease in cattle travelling from Western Australia to the Northern Territory, and in herds in the Barkly Tableland, which adjoins the Queensland border. Over the next year, the epizootic swept eastward and southward, reaching a southern limit near Sydney in early 1956. By June 1956, the epizootic had subsided but sporadic outbreaks were reported until February 1959, mainly in young cattle (Seddon 1966; Newton and Wheatley 1970). For the next 8 years, only scattered and isolated outbreaks in individual northern herds were reported.

A third major BEF epizootic was first recognised at Noonamah 43 km south of Darwin in late September 1967. It followed a similar pattern to the previous epizootics, moving progressively eastward and southward. However, apparently due to the prevailing weather conditions, this explosive epizootic moved more swiftly and extended to almost all cattle country in northern and eastern Australia (Fig. 1). During 6 weeks in early 1968, the disease moved 2000 km south along an eventual front of 800 km, reaching northern Victoria in March 1968. As in previous epizootics, mortality was low (approximately 1%) and morbidity varied from 80% in northern regions to 1%–10% in Victoria (Murray 1970).

In 1970–1971, 1972–1974 and 1974–1976, similar sweeping BEF epizootics occurred in Australia. Each appeared to originate in a region of Queensland south of the Gulf of Carpentaria. The southern sweep of the latter epizootics was interrupted during the winter months, resurging in the warmer weather from the areas most recently affected (St George et al. 1977; Uren et al. 1983). Since the late 1970s, the pattern of disease has changed from the advancing north–south wave of infection to the occurrence of isolated but sometimes extensive local outbreaks. BEF has become enzootic over a wide area of eastern Australia with a range extending from the far north to Victoria. Outbreaks now occur in the summer and autumn of each year, with successive years of high and low prevalence. Disease usually occurs in animals under 2 years of age and in older animals that escaped infection in previous years (Uren et al. 1983, 1987). The reasons for the change in the epizootiological pattern of BEF in Australia have not been determined but weather conditions favouring the successive sweeping epizootics from 1967–1976 appear to have established the virus in an enzootic ecological niche.

The pattern and distribution of BEF infection in Australia are consistent with vector-borne disease. BEF is primarily a disease of the summer and early autumn months. There is a close association between rainfall and disease outbreaks, and epidemics tend to be influenced by prevailing wind patterns (Uren 1989). In dry seasons, outbreaks are associated with streams or other ground water (Kirkland 1993). BEFV has been isolated from insects—once from *Culicoides brevitarsis*, once from a mixed pool of four mosquito species and twice from *Anopheles bancroftii* (Standfast et al. 1976; Muller and Stand-

Fig. 1 Spread of bovine ephemeral fever during the 1967–1968 epizootic in Australia. (Adapted from Murray 1970)

fast 1986). The virus has also been recovered from two species of mosquito (*Culex annulirostris*, *An. bancroftii*) and two species of biting midge (*C. brevitarsis*, *Culicoides marksi*) following experimental feeding on infected blood (Hall et al. 1975; Kay et al. 1975; St George and Standfast 1988). The abundance and distribution of the insects from which BEFV has been isolated suggests that several major vectors may be involved in transmission. The remarkable coincidence between the distribution of *Cx. annulirostris* and the geographic range of BEF also suggests that this abundant mosquito may be an important vector (Muller and Standfast 1986). The change in the epizootiological pattern of BEF since the mid-1970s may have been the result of adaptation to new vector species that has allowed local persistence of the virus between disease outbreaks. There is no evidence of direct BEFV transmission between cattle, even when nasal or ocular discharge or saliva from infected animals was smeared on mucous surfaces. However, the disease can be transmitted experimentally by injection of viraemic cattle blood (Mackerras et al. 1940; St George 1990).

4 Distribution and Epizootiology of BEF in the World

Bovine ephemeral fever is also enzootic or periodically epizootic in most tropical and sub-tropical regions of Africa, the Middle-East and Asia (Odend'hal 1983; St George 1988). The disease has been reported in many countries in Africa, but detailed information on the epizootiology has been reported primarily from Zimbabwe (Bevan 1907, 1912), Nigeria (Kemp et al. 1973b), Kenya (Davies et al. 1975; Davies and Walker 1974; Davies et al. 1990) and South Africa (Freer 1910; Macfarlane and Haig 1955; van der Westhuizen 1967). In tropical regions of central Africa, BEF is enzootic. Epizootics have been reported to occur seasonally, sweeping north or south of the Equator into more temperate zones (St George 1988). The infection is enzootic in southern Egypt and has extended in seasonal epizootics into Israel, Iran, Iraq, Kuwait, Jordan, Lebanon, Saudi Arabia and Turkey. It has not been reported in countries of the northern or western Mediterranean but there is serological evidence of infection in southern Russia (Kurchenko et al. 1991; Davies et al. 1993). BEF has never been reported in the Americas, in New Zealand or islands of the Pacific, where it is considered an important exotic disease.

In the eastern hemisphere, BEF occurs in most Asian countries south of the 38°N parallel, extending from Afghanistan to Japan in the north-east, and south to Irian Jaya and Papua New Guinea (St George 1990). In Indonesia, the virus is enzootic, infections are widespread and disease outbreaks are common, particularly during the wet season (Daniels et al. 1993). In China, BEF is enzootic in the southern tropical provinces and occurs regularly in most parts of the country. In Guangdong Province, epizootics have occurred at 1- to 2-year intervals, often extending to central provinces (Bai et al. 1991; Bai 1993). In 1983 and 1991, sweeping epidemics reached Liaoning and Jilin Provinces, the most northern recorded extent of BEF (44°N). In East Asia, BEF epizootics also occur in Taiwan, South Korea and Japan. There appears to be a temporal relationship between some outbreaks in South Korea and Japan (Shirakawa et al. 1994), and these may have their origins in the enzootic regions of southern China. In Taiwan, epizootics have occurred in 1967, 1983, 1989, 1996 and 1999 (Wang et al. 2001). As in Australia, the pattern of disease in Taiwan has evolved from sweeping epizootic to slow epizootic and, most recently, to compact local outbreaks (Liao et al. 1998; Wang et al. 2001).

As in Australia, BEF in Africa and Asia usually occurs in the summer and autumn months, and is often associated with periods of high rainfall. The spread of epizootics appears to follow the pattern of prevailing winds, with a general northward sweep above the equator and a southerly movement in the southern hemisphere (St George 1988). The virus has been isolated from biting midges (mixed pool of *Culicoides* spp.) in Kenya and Zimbabwe (*Culicoides coarctatus*) but no other insect isolations have been reported outside of Australia (Davies and Walker 1974; Blackburn et al. 1985; Venter et al. 2003). Vertebrate hosts include cattle (*Bos taurus, Bos indicus, Bos javanicus*) and water buffaloes (*Bubalus bubalis*), in which infection can cause severe to mild disease or may be sub-clinical. There is also serological evidence of subclinical infection in a wide range of wild ruminant species in Africa (St George 1990; Barnard 1997; Anderson and Rowe 1998). Sheep can be infected experimentally with no clinical signs of disease, but there is no evidence of natural infection of sheep in Australia (Hall et al. 1975).

5 Pathogenesis of BEFV Infection

The available evidence indicates that the clinical signs of BEFV infection are due primarily to a vascular inflammatory response. In experimental infections, the incubation period is usually only from 2 to 4 days, and viraemia and clinical signs are of short duration, persisting for only 1–3 days (Mackerras et al. 1940; St. George et al. 1984; Uren et al. 1989, 1992). There is no evidence of extensive tissue damage other than a vasculitis affecting the endothelium of small vessel synovial membranes, tendon sheaths, muscles, fascia and skin (Mackerras et al. 1940; Basson et al. 1969; Kirkland 2002). Although the initial site of infection following transmission is not known, the virus is subsequently detected primarily in leucocytes and blood plasma (Mackerras et al. 1940; Young and Spradbrow 1980; St George 1993). The onset of fever and other clinical signs is accompanied by a marked leucopoenia, relative neutrophilia, elevated plasma fibrinogen, biochemical imbalance including hypocalcaemia, and elevated levels of cytokines (Uren and Murphy 1985; St George et al. 1986; Uren et al. 1989; Uren et al. 1992). There are some reports of prolonged paralysis or ataxia in animals recovering from the acute effects of infection (Mackerras et al. 1940; Macfarlane and Haig 1955; Basson et al. 1970; Snowdon 1970). However, severe symmetrical Wallerian degeneration observed in the spinal cords is considered to be the result of trauma associated with pressure on the chord (Hill and Schultz 1977). The major clinical signs of fever, anorexia, muscle stiffness, ocular and nasal discharge, rumenal stasis and sternal recumbency can be very effectively treated with anti-inflammatory drugs without affecting the development of neutralising antibodies or resistance to subsequent challenge (Uren et al. 1989). Interferon toxicity has been suggested as the cause of the general inflammatory response and other clinical signs of BEFV infection (St George 1993).

6
BEFV Morphology and Structure

Bovine ephemeral fever virus (BEFV) is the type species of the genus *Ephemerovirus* in the family *Rhabdoviridae* (Walker et al. 1999). BEFV virions are approximately 70×180-nm bullet- or cone-shaped, enveloped particles that are typical of animal rhabdoviruses (Fig. 2) (Ito et al. 1968; Murphy et al. 1972; Della-Porta and Snowdon 1980). Virions contain a helical nucleocapsid, comprising the (–) RNA genome, a protective nucleoprotein (N), and the large (L) and small (P) subunits of an RNA-dependent RNA polymerase. The ribonucleoprotein (RNP) complex, comprising the RNA, N, L and P proteins, is the active replication and transcription unit of the virus. The structural proteins also include a matrix protein (M) and a class I transmembrane glycoprotein (G), each of which appear to be phosphorylated (Walker et al. 1991). The 81-kDa G protein spans the viral envelope to form clear projections on the virion surface. It binds to cells to facilitate invasion by receptor-mediated endocytosis, is the target for type-specific and neutralizing antibody (Walker et al. 1991; Cybinski et al. 1990) and protects cattle against experimental BEFV infection (Uren et al. 1994; Hertig et al. 1996).

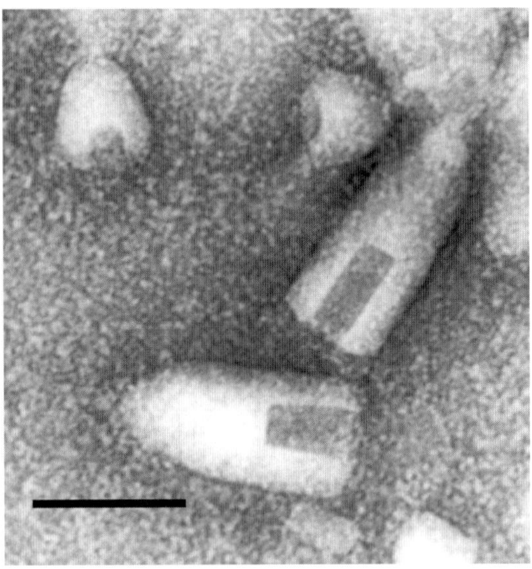

Fig. 2 BEFV virions and defective-interfering (*DI*) particles. The bar represents 100 nm. (Reproduced with permission from Springer, Heidelberg)

The BEFV virion can be disrupted progressively by treatment with detergent in the presence of increasing salt concentrations, generating relatively pure preparations of the major structural proteins (Walker et al. 1991; Riding et al. 1993). BEFV preparations often also include truncated, defective-interfering (DI) particles that contain incomplete genomes and cannot replicate in the absence of infectious virus (Della-Porta and Snowdon 1980).

7
BEFV Genome Organization and Expression

Bovine ephemeral fever virus has the largest and most complex of known rhabdovirus genomes. The genome structure, comprising 14,900 nucleotides of (–) sense single-stranded RNA, is illustrated in Fig. 3 (GenBank Accession No. NC 002526). As for other rhabdoviruses, there are five structural protein genes (N, P, M, G and L) flanked by partially complementary, 3′ and 5′ non-coding termini. However, between the G and L genes, BEFV contains a complex arrangement of genes that encode several proteins that are unique to ephemeroviruses. These include a second non-structural glycoprotein (G_{NS}) and five open reading frames (ORFs) that encode putative smaller proteins ($\alpha1$, $\alpha2$, $\alpha3$, β and γ) of unknown function (Walker et al. 1992; McWilliam et al. 1997). Additional small ORFs occur in alternative frames in the P and $\alpha2$ genes, but it is not known if the products are expressed (Walker et al. 1999).

The 90-kDa G_{NS} glycoprotein is a class I transmembrane glycoprotein that is related in structure and sequence to the BEFV G protein and to other rhabdovirus G proteins, and appears to have been generated by gene duplication (Walker et al. 1992; Wang and Walker 1993). The G_{NS} protein is abundant in BEFV-infected cells but it is not present in virions. It does not share antigenic

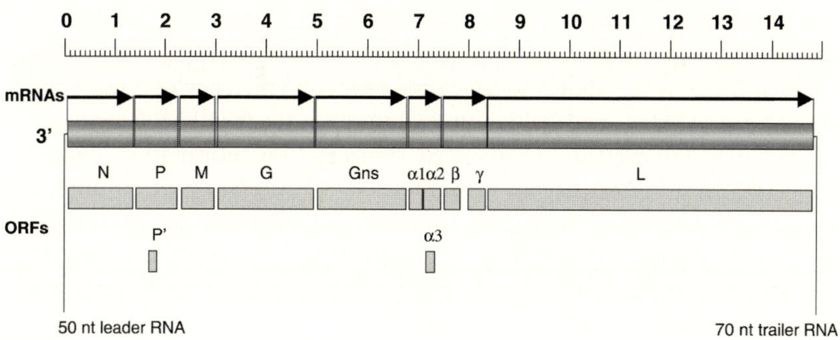

Fig. 3 Organization of the BEFV genome. (Reproduced with permission from Springer, Heidelberg)

sites with the G protein, and G_{NS} antibody does not neutralize the virus produced in mammalian or insect cells (Hertig et al. 1996). The function of the G_{NS} protein requires further study.

The proteins encoded in the $\alpha 1$, $\alpha 2$, $\alpha 3$, β and γ ORFs have yet to be identified in BEFV-infected cells. The $\alpha 1$ protein has been expressed in *Escherichia coli* and in insect cells from a recombinant baculovirus vector, and it has been shown to be cytotoxic (Dhillon 1996). The deduced amino acid sequence of the $\alpha 1$ protein features a transmembrane domain and highly basic C-terminal region, suggesting that it functions as a viroporin (McWilliam et al. 1997). The functions of the other novel BEFV proteins remain unclear but each is preserved in Australian (BB7721/1968) and Chinese (Beijing-1/1976) BEFV isolates, indicating that they are essential to efficient viral replication (McWilliam et al. 1997). The G_{NS}, $\alpha 1$, $\alpha 2$ and β protein genes also occur in the related ephemerovirus *Adelaide River virus* which lacks the $\alpha 3$ and γ genes (Wang and Walker 1993; Wang et al. 1994). There is some evidence that the ARV β protein may be a virion component (Wang and Walker 1993).

As for other rhabdoviruses, transcription of polyadenylated BEFV mRNAs initiates and terminates at short conserved sequences flanking each coding region. The N, P, M, G, G_{NS} and L genes initiate from a UUGUCC sequence (mRNA: 5′-cap-AACAGG...) and terminate at GUAC$(U)_7$-3′, the putative polyadenylation site. Transcription from the G_{NS}-L intergenic region, encompassing the $\alpha 1$, $\alpha 2$, $\alpha 3$, β and γ ORFs, appears to be more complex (McWilliam et al. 1997). The primary transcripts are polycistronic α ($\alpha 1$-$\alpha 2$-$\alpha 3$) and β-γ mRNAs. However, the α coding region is flanked by the conserved UUGUCC initiation sequence and the variant GU$\underline{U}$C$(U)_7$ sequence, which appears to cause incomplete transcription termination, resulting in leaky expression of a longer α-β-γ mRNA. The β-γ coding region is flanked by the highly conserved UUGUCC and GUAC$(U)_7$ transcription regulation sequences, but the termination sequence overlaps the downstream L gene by 21 nucleotides. This type of overlap also occurs in ARV (Wang et al. 1994) and in the pneumovirus respiratory syncytial virus (Collins et al. 1987) in which it was shown to have little effect on efficiency of transcription of the L gene (Fearns and Collins 1999). Interestingly, the BEFV β gene is also immediately followed by a truncated transcription termination sequence GUAC$(U)_6$ and a downstream conserved UUGUCC transcription initiation sequence in advance of the γ gene. However, as shown for VSV, truncation of the U_7 palindrome appears to result in ineffective transcription termination and polyadenylation (Barr et al. 1997), precluding initiation of transcription of a monocistronic γ-mRNA.

8
BEF-Related Viruses

In addition to BEFV, the genus *Ephemerovirus* also includes *Berrimah virus* (BRMV), *Adelaide River virus* (ARV) and tentatively, *Kimberley virus* (KIMV), *Malakal virus* (MALV) and *Puchong virus* (PUCV) (Walker et al. 1999). BRMV was isolated only once from a healthy sentinel steer near Darwin in 1981 (Gard et al. 1982). It is closely related to BEFV, sharing several of the major neutralization sites (Cybinski et al. 1990, 1992). *Kimberley virus* was first isolated from *Cx. annulirostris* collected in the Ord River region of Western Australia in 1973 (Liehne et al. 1981) and has been isolated subsequently from *C. brevitarsis* and from healthy cattle (Cybinski and Zakrzewski 1983; Zakrzewski and Cybinski 1984). *Adelaide River virus* was first isolated from a healthy sentinel steer in the Northern Territory in 1981 (Gard et al. 1984). There is serological evidence that BRMV, KIMV and ARV are present in China and Indonesia (Bai et al. 1993; Daniels et al. 1995). MALV and PUCV have been isolated from *Mansonia uniformis* mosquitoes in Sudan and western Malaysia, respectively. Each of these viruses cross-reacts strongly with BEFV in IFA tests but can be distinguished by neutralization tests. However, experimental or natural infection with some of these viruses can produce heterotypic neutralizing antibody, potentially confusing serological diagnosis of infection (Cybinski 1987). None of the BEF-related viruses is known to cause disease in cattle. Interestingly, *Kotonkan virus*, which was originally isolated from a mixed pool of *Culicoides* spp. in Nigeria, does cause an ephemeral fever-like illness in cattle (Kemp et al. 1973a; Tomori et al. 1974, 1975). Indeed, although *Kotonkan virus* was previously regarded to be related to *Mokola virus* and other lyssaviruses (Kemp et al. 1973a), phylogenetic analysis of nucleotide sequences in the polymerase (L) gene has indicated a close relationship to the ephemeroviruses (H. Bouhry, J.A. Cowley, P.J. Walker, unpublished data).

9
Antigenic Structure of BEFV

The major BEFV antigens are the G and N proteins, each of which occurs abundantly in virions and in infected cells (Walker et al. 1991). The G protein contains type-specific and neutralizing antigenic sites and induces protective immunity in cattle (Cybinski et al. 1990; Uren et al. 1994). The N protein is the major cross-reactive BEFV antigen which is detected in group-specific complement fixation (CF) and indirect fluorescent antibody (IFA) tests. The

Rabies virus N protein can protect against experimental infection (Dietzschold et al. 1987), but there is no evidence that the BEFV N protein has protective properties.

The antigenic structure of the BEFV G protein has been examined in some detail (Fig. 4). Competitive-binding assays using a panel of G protein monoclonal antibodies and analysis of the cross-neutralization of monoclonal antibody-selected escape mutants has identified four distinct neutralization sites (G1, G2, G3 and G4) (Cybinski et al. 1990 1992). G1 is a linear site located at amino acids 487–503 in the stem domain of the 623 amino acid G protein. The site comprises two minimal B cell epitopes that map to each end of the antigenic domain. G2 is a conformational site that is adjacent to a putative disulphide bridge connecting a tight loop in the G protein at amino acids 172–182. G3 is comprised of two partially overlapping elements (G3a and G3b). It is the major conformational site, combining complex discontinuous epitopes derived from distant locations in the cysteine-rich "head" structure of the G protein (Kongsuwan et al. 1998). Corresponding, but non-cross-reacting, conformational sites have been identified in the G proteins of other animal rhabdoviruses (Walker and Kongsuwan 1999). G4 is a linear site defined by a single monoclonal antibody which cross-neutralizes *Berrimah virus* and *Kimberley virus* (Cybinski et al. 1992). Site G4 has not yet been located in the G protein structure.

The antigenic structure of the BEFV N protein has also been examined, but in less detail than the G protein (Walker et al. 1994). All N protein monoclonal antibodies obtained to date have been non-neutralizing and directed at non-conformational epitopes. Each reacts with both a full-length recombinant N protein expressed in *E. coli* as a GST-fusion protein, and a 244 amino acid C-terminal fragment of the N protein, but not with the complementary N-terminal fragment. All N monoclonal antibodies cross-react in IFA tests with *Berrimah virus* and one reacts with *Kimberley virus*. Two BEFV N protein monoclonal antibodies have also been shown to react with baculovirus-expressed *Rabies virus* N protein. It has also been shown that polyclonal antisera to BEFV and ARV cross-react with *Rabies virus* N protein (Walker et al. 1994) and, in IFA tests, extensive low-level cross-reactions have been reported between ephemeroviruses, lyssaviruses and several unclassified rhabdoviruses isolated from cattle or insects (Calisher et al. 1989). However, an analysis of N protein amino acid sequences indicates few regions of linear conservation and, overall, ephemerovirus N protein sequences are more closely related to vesiculoviruses than to lyssaviruses. This is also true of other structural proteins (L and G), and it is clear that serological links between lyssaviruses and ephemeroviruses are determined

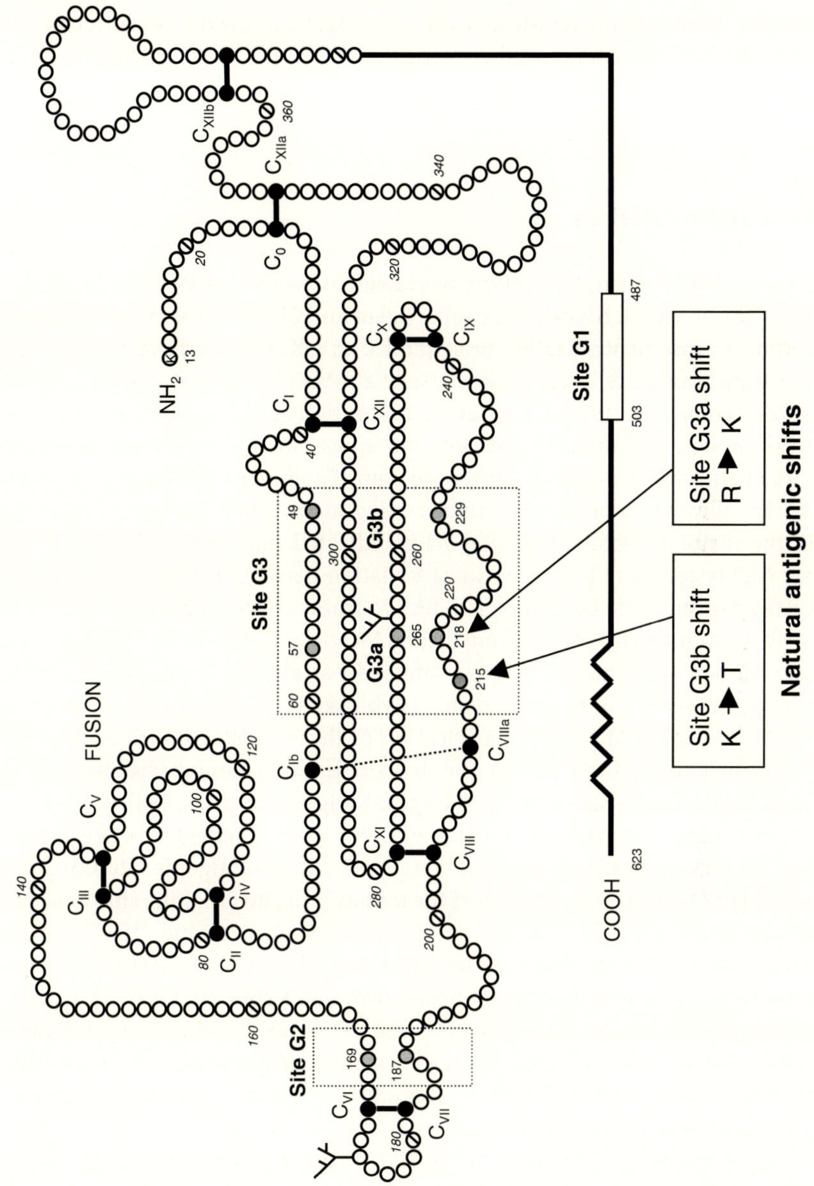

Fig. 4 Deduced structural model of the BEFV G protein (Walker and Kongsuwan 1999) showing sites of natural and experimental antigenic variation in neutralization site G3

by very limited similarities that do not reflect the overall evolutionary or ecological relationships of the viruses (Walker et al. 1994; Dhillon et al. 2000).

10 BEFV Strain Variation

BEFV exists as a single serotype worldwide. Comparisons of BEFV isolates from South Africa, Nigeria, Australia, Japan and China by cross-protection in cattle, or cross-neutralization tests in mice or cell cultures, have not revealed any significant differences (Takematsu et al. 1956; Inaba et al. 1969; Snowdon 1970; Kemp et al. 1973b; Tian et al. 1987), probably because viraemia and transmission of infection precedes the appearance of significant levels of virus neutralising antibody by several days (Snowdon 1970; Uren et al. 1989, 1992). However, more recent studies using monoclonal antibodies directed at the major neutralization sites have indicated that, although G1 and G2 are highly conserved, there is some variation between isolates in the major conformational site G3. Among 18 BEFV strains collected between 1956 and 1989 from cattle and insects from distant locations in Australia, the prototype structure of antigenic site G3a (as defined by strain BB7721 isolated in 1968) was preserved in only five isolates (Cybinski et al. 1992). An analysis of the amino acid sequences of a selection of these isolates indicated that the antigenic shift in site G3a was due to a single change in a basic amino acid (arginine to lysine) at residue 218 (Fig. 4; Kongsuwan et al. 1998).

An expanded study of 70 Australian BEFV isolates collected between 1956 and 1992 has identified a second antigenic shift in antigenic site G3b. Nucleotide sequence analysis of 20 of the strains indicated that the shift was due to a single amino acid change (lysine to threonine) at residue 215 (Fig. 4; P.J. Walker and K. Kongsuwan, unpublished data). It was also observed that the shift in site G3a was evident in most isolates collected after 1973, but there was not a clear temporal basis for the shift in site G3b. As the epizootiological pattern of BEFV in Australia changed dramatically in the late 1970s, it would be of interest to determine if the shift in site G3a is associated with an adaptive change in vector competence.

Comparison of Australian (BB7721/1968) and Chinese (Beijing-1/1976) strains has also indicated that all sites other than G3a are preserved. Comparison of the G protein amino acid sequences of BEFV isolates from Australia and Taiwan has indicated the Taiwanese strain Tn88128/1999 is more distantly related than any of the available Australian isolates. The Taiwanese strain is most closely related to the first insect isolate of BEFV (CS53/1974), which was obtained from a mixed pool of mosquitoes collected in central Queensland in April 1974 (Fig. 5). A comparison of the G proteins of six Taiwanese strains isolated between 1989 and 1999 has indicated closer amino acid sequence identity to Japanese than to Australian strains, and significant variations sequence from the prototype Australian BB7721/1968 in antigenic sites G1 and G3 (Wang et al. 2001). The patterns of amino acid substitution in the Taiwanese isolates indicated that they fell into two clusters, the first of which included the 1984 vaccine strain and the second of which appears to have emerged since 1996 (Wang et al. 2001).

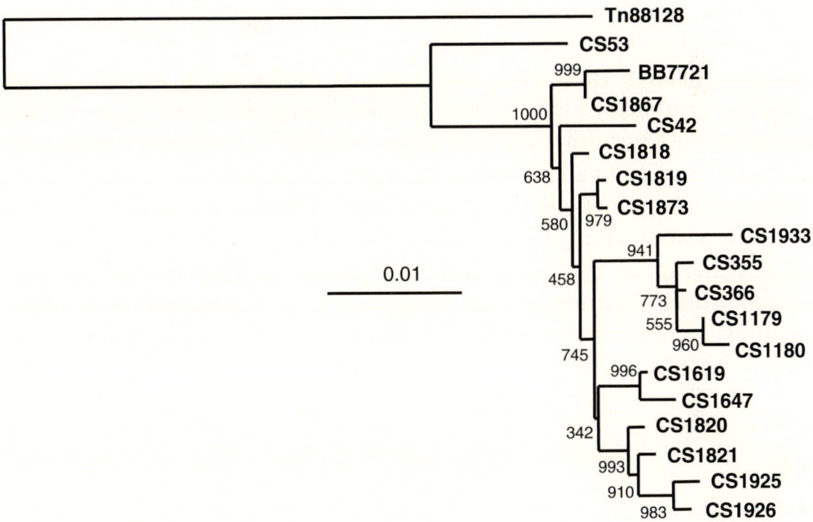

Fig. 5 Phylogenetic analysis of G gene sequences of 18 Australian BEFV isolates and Taiwanese isolate Tn88128 (GenBank AF208840). The tree was constructed from a Clustal W multiple sequence alignment using the neighbour-joining method. Tn88128 was selected as the outgroup. Bootstrap values were estimated from 1,000 trees. Branch lengths are proportional to distance. The bar represents a sequence divergence of 1%. Nucleotide sequence analysis of Australian BEFV isolates was conducted by Dr. K. Kongsuwan (CSIRO, Australia). Origin of isolates: Tn88128 Kaoshiung/Tw/1999; CS53 Etna Creek/Au/1974; BB7721 Charters Towers/Au/1968; CS1867 Etna Creek/Au/1970; CS42 Beatrice Hill/Au/1975; CS1818 Upper Barron/Au/1970; CS1819 Wacol/Au/1973; CS1873 Etna Creek/Au/1970; CS1933 Etna Creek/Au/1973; CS355 Peachester/Au/1980; CS366 Peachester/Au/1980; CS1179 Peachester/Au/1982; CS1180 Peachester/Au/1982; CS1619 Peachester/Au/1984; Peachester/Au/1984; CS1820 Peachester/Au/1976; CS1821 Amberley/Au/1975; CS1925 Tolga/Au/1975; CS1926 Kairi/Au/1975

11 Disease Diagnosis

During the course of epidemics or local disease outbreaks, ephemeral fever diagnosis is usually based on the rapid onset and recovery from the typical clinical signs (St George 1990; Kirkland 2002). However, early sporadic cases in disease outbreaks or isolated cases in areas where the disease is now known to occur can be difficult to distinguish from hemoprotozoan or respiratory viral infections (Bai et al. 1987; St. George 1988). Laboratory diagnosis should indicate rising titres of neutralizing antibody in paired acute and convalescent sera by using a virus neutralization test (Tzipori 1975; St George et al. 1980; Cybinski 1987) or blocking ELISA (Zakrzewski et al. 1992). A complement fixation test has also been used for confirmatory diagnosis (Bai et al. 1987). However, serological diagnosis can be complicated by prior infection with other ephemeroviruses and should be interpreted cautiously in the absence of typical clinical signs (Cybinski 1987).

Virus isolation from cattle blood can be conducted in hamster kidney (BHK21) or monkey kidney (Vero) cell lines (Snowdon 1970; Tzipori 1975). However, the short duration of viraemia in cattle and inefficient replication of the virus prior to adaptation to mammalian cell cultures has limited the usefulness of this method. Direct inoculation of mosquito (*Ae. albopictis* C6–36) cell lines and detection by indirect immunofluorescence prior to passage in BHK21 cells appears to be a more effective procedure (Uren 1983). PCR-based detection of virus during the acute phase of infection is now commonly used in Australia for rapid laboratory diagnosis (B. Corney, personal communication; Kirkland 2002).

12 BEF Vaccines

Natural BEFV infection induces a strong neutralizing antibody response and, although some exceptions have been recorded, infection appears to induce a durable immunity (Mackerras et al. 1940). It is also known that colostral antibody will protect calves against experimental challenge (St George 1986) but there are conflicting reports of the role of neutralizing antibody in protection (Della-Porta and Snowdon 1979b; Uren et al. 1994). Although a T-cell response to BEFV antigens has been observed (Uren et al. 1993), the relative importance of humoral and cell-mediated immunity in protection is unclear.

Several forms of live-attenuated, inactivated, subunit and recombinant BEF vaccines have been reported. The efficacy and longevity of protection has been variable. Vaccines manufactured in South Africa, Japan, China, Taiwan and Australia have been used in the field for BEF prophylaxis. Live-attenuated vaccines have been produced by serial passage of BEFV in suckling mice and/or in cell cultures (Tzipori and Spradbrow 1973, 1978; Theodoridis et al. 1973a; Spradbrow 1975; Vanselow et al. 1995). In general, live vaccines have been administered in combination with adjuvants (including Freund's complete or incomplete adjuvant, dextran sulphate, aluminium hydroxide or Quil A). It has been reported that Quil A and aluminium hydroxide gel adjuvants inactivate up to 99% of BEFV infectivity (Vanselow et al. 1985). However, it is also known that Quil A can cause aggregation of BEFV particles and that, while the apparent infectivity of high-titre BEFV preparations was clearly reduced following combination with Quil A, the effect is due to aggregation rather than inactivation (P.J. Walker, unpublished data). Live-attenuated vaccines appear to be effective in inducing immunity but require at least two doses in adjuvant to stimulate durable protection.

Inactivated BEFV vaccines have also been reported. Two doses of formalin-inactivated BEFV have been shown to protect calves against experimental bovine infection, but protection was not as effective as a vaccine employing consecutive live-attenuated and killed virus inoculations (Inaba et al. 1973, 1974). A formalin-inactivated BEFV vaccine was reported to be ineffective in inducing immunity and variable protection was obtained using virus inactivated with 2-propiolactone (Della-Porta and Snowdon 1979a, 1979b). However, there was no correlation between neutralizing antibody titre and protection, and a number of animals with high neutralizing antibody titres did not resist challenge.

More recently, the G protein subunit purified from BEFV virions has been used in conjunction with Quil A adjuvant (Uren et al. 1994). The vaccine gave consistent protection in all vaccinated animals when the vaccine dose was at or above 0.8 µg of protein. There was a strong neutralizing antibody response to both conformational and non-conformational antigenic sites in all protected animals and, in two animals vaccinated with the sub-protective dose (0.08 µg/dose), the neutralizing antibody response to vaccination was poor. A crude Triton X-100-solubilized extract of BEFV-infected cell cultures has also been shown to be protective when administered in two doses with Freund's complete adjuvant or white oil adjuvant. The BEFV G protein is likely to be the active component of this vaccine (Bai et al. 1993). A recombinant vaccinia virus vector expressing the BEFV G protein has also been reported to induce protective immunity after a single vaccination in the absence of adjuvant. However, vaccinia virus expressing recombinant BEFV G_{NS} protein failed to induce virus-neutralising antibodies or protect against experimental challenge (Hertig et al. 1996).

13
Conclusion

Bovine ephemeral fever remains an important disease of livestock and further research is required to resolve several important issues relating to the virus and its ecology. Although both mosquitoes and biting midges are known to carry the virus, the enzootic and epizootic ecology have not been clearly established anywhere in the world. The major vector species have not been determined and the mechanisms of over-wintering and persistence between BEFV outbreaks are not known. Major changes in the epizootiological pattern, as has occurred in Australia, may have arisen through viral adaptation to new vectors, allowing the establishment of sustainable disease cycles in sub-tropical and temperate climes. Molecular analysis of BEFV isolates, in conjunction with studies of vector competence, will assist in understanding the ecology and epizootiology of BEFV infection.

There is also a need for a better understanding of the role of BEF-related viruses in disease. There have been reports of BEF-like disease in cattle with no serological evidence of BEFV infection, examples of multiple disease episodes in the same cattle during single or successive years, and reports of BEF vaccine breakdown are common. Although there is no evidence that some of the known BEF-related viruses cause disease, *Kotonkan virus* in Africa does cause a BEF-like illness, and more information is required on the host range and pathogenicity of *Malakal virus* in Africa and *Puchong virus* in Asia. It is also likely that other BEF-related viruses have yet to be identified.

Vaccination now has an important role in the management of bovine ephemeral fever. However, vaccines currently in use are less than ideal, and recent technological advances offer opportunities for the development of new heat-stable vaccines that will provide reliable and durable protection following a single dose. The BEFV G gene is now recognised as a potent vaccine antigen and several recombinant vaccines employing G gene clones are currently under development. Recombinant vaccines based on viral vectors or naked DNA also offer the prospect of multivalent antigen delivery in conjunction immuno-regulatory proteins to enhance efficacy and durability of protection.

References

Anderson EC, Rowe LW (1998) Prevalence of antibody to bovine virus diarrhoea, bovine herpes virus 1, rift valley fever, ephemeral fever and bluetongue and to *Leptospira* sp. in free ranging wildlife in Zimbabwe. Epidemiol Infect 121:441–449

Bai W (1993) Epidemiology and control of bovine ephemeral fever in China. In: St. George TD, Uren MF, Young PL, Hoffman D (eds) Bovine ephemeral fever and related rhabdoviruses. ACIAR, Canberra, pp 20–22

Bai W, Tian F, Wang C, Jiang C, Zhang Z (1987) Preliminary studies of the complement fixation test to confirm the diagnosis of bovine ephemeral fever. Aust J Biol Sci 40:137–141

Bai W, Jiang C, Davis SS (1991) Preliminary observations on the epidemiology of bovine ephemeral fever in China. Trop Anim Hlth Prod 23:22–26

Bai W, Yan J, Zhang Z, Jiang C and Lin Y (1993) Studies on a vaccine against ephemeral fever. In: St. George TD, Uren MF, Young PL, Hoffman D (eds) Bovine ephemeral fever and related rhabdoviruses. ACIAR, Canberra, pp 111–114

Barnard BJ (1997) Antibodies against some viruses of domestic animals in southern African wild animals. Onderstepoort J Vet Res 64:95–110

Barr JN, Whelan SPJ, Wertz GW (1997) *Cis*-acting signals involved in termination of vesicular stomatitis virus mRNA synthesis include the conserved AUAC and the U7 signal for polyadenylation. J Virol 71:8718–8725

Basson PA, Pienaar JG, van der Westhuizen B (1969) The pathology of ephemeral fever: a study of experimental disease in cattle. J Sth Afr Vet Med Assoc 40:385–397

Bevan LEW (1907) Preliminary report on the so-called "stiff sickness" or "three-day-sickness" of cattle in Rhodesia. J Com Pathol Ther 20:104–113

Bevan LEW (1912) Ephemeral fever, or three day sickness in cattle. Vet J 68:458–461

Blackburn NK, Searle L, Phleps RJ (1985) Viruses isolated from *Culicoides* (Diptera: *Ceratopogonidae*) caught at the veterinary research farm, Mazowe, Zimbabwe. J Entomol Soc Sth Afr 48:331–336

Calisher CH, Karabatsos N, Zeller H, Digoutte J-P, Tesh RB, Shope RE, Travassos da Rosa APA, St. George TD (1989) Antigenic relationships among rhabdoviruses from vertebrates and haematophagous arthropods. Intervirology 50:241–257

Chenoweth PJ, Burgess GW (1972) Mid-piece abnormalities in bovine semen following ephemeral fever. Aust Vet J 48:37–38

Cybinski DH (1987) Homologous and heterologous antibody reactions in sera from cattle naturally infected with bovine ephemeral fever group viruses. Vet Microbiol 13:1–9

Cybinski DH, Zakrzewski H (1983) The isolation and preliminary characterization of a rhabdovirus in Australia related to bovine ephemeral fever virus. Vet Microbiol 8:221–235

Cybinski DH, Walker PJ, Byrne KA, Zakrzewski H (1990) Mapping of antigenic sites on the bovine ephemeral fever virus glycoprotein using monoclonal antibodies. J Gen Virol 71:2065–2072

Cybinski DH, Davis SS, Zakrzewski H (1992) Antigenic variation of the bovine ephemeral fever virus glycoprotein. Arch Virol 124:211–224

Daniels PW, Sendow I, Soleha E, Sukarhih, Hunt NT, Bahri S (1995) Australian-Indonesian collaboration in veterinary arbovirology – a review. Vet Microbiol 46:151–174

Daniels PW, Soleha E, Sendow I, Sukarsih (1993) Bovinc ephemeral fever in Indonesia. In "Bovine Ephemeral Fever and Related Rhabdoviruses" (St. George TD, Uren MF, Young PL, Hoffman D, eds.), pp 41–44. ACIAR, Canberra

Davies FG, Moussa A, Barsouma G (1993) The 1990–1991 epidemic of ephemeral fever in Egypt and the potential for spread to the Mediterranean region. In: St. George TD, Uren MF, Young PL, Hoffman D (eds) Bovine ephemeral fever and related rhabdoviruses. ACIAR, Canberra, pp 54–56

Davies FG, Ochieng P, Walker AR (1990) The occurrence of ephemeral fever in Kenya, 1968–1988. Vet Microbiol 22:129–136

Davies FG, Shaw T, Ochieng P (1975) Observations on the epidemiology of ephemeral fever in Kenya. J Hyg 75:231–235

Davies FG, Walker AR (1974) The isolation of ephemeral fever virus from cattle and *Culicoides* midges in Kenya. Vet Rec 95:63–64

Davis SS, Gibson DS, Clark R (1984) The effect of bovine ephemeral fever on milk production. Aust Vet J 61:128–129

Della-Porta AJ, Snowdon WA (1979a) An experimental inactivated virus vaccine against bovine ephemeral fever 1. Studies of the virus. Vet Microbiol 4:183–195

Della-Porta AJ, Snowdon WA (1979b) An experimental inactivated virus vaccine against bovine ephemeral fever 2. Do neutralizing antibodies protect against infection? Vet Microbiol 4:197–208

Della-Porta AJ, Snowdon WA (1980) Bovine ephemeral fever virus. In: Bishop DHL (ed) Rhabdoviruses, Vol 3. CRC Press, Boca Raton, pp 167–191

Dhillon J (1996) Molecular study of bovine ephemeral fever virus proteins. PhD thesis, The University of Queensland, Australia

Dhillon J, Cowley JA, Wang YH, Walker PJ (2000) RNA polymerase (L) gene and genome terminal sequences of ephemeroviruses *Bovine ephemeral fever virus* and *Adelaide River virus* indicate a close relationship to vesiculoviruses. Virus Res 70:87–95

Dietzschold B, Wang H, Rupprecht CE, Celis E, Tollis M, Ertl H, Heber-Katz E, Koprowski H (1987) Induction of protective immunity against rabies by immunization with the rabies virus ribonucleoprotein. Proc Natl Acad Sci USA 84:9165–9169

Fearns R, Collins PL (1999) Model for polymerase access to the overlapped L gene of respiratory syncytial virus. J Virol 73:388–397

Freer GW (1910) Ephemeral fever or three-day sickness of cattle. Vet J 66:19–22

Gard GP, Cybinski DH, St. George TD (1982) The isolation in Australia of a new virus related to bovine ephemeral fever virus. Aust Vet J 60:89

Gard GP, Cybinski DH, Zakrzewski H (1984) The isolation of a fourth bovine ephemeral fever group virus. Aust Vet J 61:332

Hall WT, Daddow KN, Dimmock CK, St. George TD, Standfast HA (1975) The infection of merino sheep with bovine ephemeral fever virus. Aust Vet J 51:344

Hertig C, Pye AD, Hyatt AD, Davis SS, McWilliam SM, Heine HG, Walker PJ, Boyle DB (1996) Vaccinia virus-expressed bovine ephemeral fever virus G but not GNS glycoprotein induces neutralizing antibodies and protects against experimental infection. J Gen Virol 77:631–640

Hill MWM, Schultz K (1977) Ataxia and paralysis associated with bovine ephemeral fever infection. Aust Vet J 53:217–221

Inaba Y, Sato K, Tanaka Y, Ito H, Omori T, Matumoto M (1969) Serological identification of bovine epizootic fever virus as ephemeral fever virus. Jap J Microbiol 13:388–389

Inaba Y, Kurogi H, Sato K, Goto Y, Omori T, Matumoto M (1973) Formalin-inactivated, aluminium phosphate gel-adsorbed vaccine of bovine ephemeral fever virus. Arch Ges Virusforsch 44:42–53
Inaba Y, Kurogi H, Takahashi A, Sato K, Omori T, Goto Y, Hanaki T, Yamamoto M, Kishi S, Kodama K, Harada K, Matumoto M (1974) Vaccination of cattle against bovine ephemeral fever with live attenuated virus followed by killed virus. Arch Ges Virusforsch 44:121–132
Ito Y, Tanaka Y, Inaba Y, Omori T (1968) Electron microscopic observations of bovine enzootic fever virus. Natl Inst Anim Hlth Quart 9:35–44
Kay BH, Carley JG, Filippich C (1975) The multiplication of Queensland and New Guinean arboviruses in *Culex annulirostris* (Skuse) (Diptera: Culicidae) J Med Entomol 12:279–283
Kemp GE, Lee VH, Moore VL, Shope RE, Causey OR, Murphy FA (1973a) Kotonkan, a new rhabdovirus related to Mokola virus of the rabies serogroup. Am J Epidemiol 98:43–49
Kemp GE, Mann ED, Tomori O, Fabiyi A, O'Connor E (1973b) Isolation of bovine ephemeral fever virus in Nigeria. Vet Rec 93:107–108
Kirkland PD (2002) Akabane and bovine ephemeral fever virus infections. Vet Clin Food Admin 18:501–514
Kirkland PD (1993) The epidemiology of bovine ephemeral fever in south-eastern Australia: evidence for a mosquito vector. In: St. George TD, Uren MF, Young PL, Hoffman D (eds) Bovine ephemeral fever and related rhabdoviruses. ACIAR, Canberra, pp 33–37
Kongsuwan K, Cybinski DH, Cooper J, Walker PJ (1998) Location of neutralizing epitopes on the G protein of bovine ephemeral fever rhabdovirus. J Gen Virol 79:2573–2518
Kurchenko FP, Gononov YM, Khlybova TV, Zaytzev VL, Kekukh IG, Pasechnikov LN, Alekhin AF, Vyatkina NV (1991) Recovery and identification of the cattle virus ephemeral fever. Veterinariya (Moskva) 2:26–28
Liao YK, Inaba Y, Li NJ, Chain CY, Lee SL, Liou PP (1998) Epidemiology of bovine ephemeral fever virus infection in Taiwan. Microbiol Res 153:289–295
Liehne PFS, Anderson S, Stanley NF, Liehne CG, Wright AE, Chan KH, Leivers S, Britten DK, Hamilton NP (1981) Isolation of Murray Valley encephalitis virus and other arboviruses in the Ord River valley in 1972–1976. Aust J Exp Biol Med Sci 59:347–356.
Macfarlane IS, Haig DA (1955) Some observations on three-day stiff-sickness in the Transvaal in 1954. J Sth Afr Vet Med Assoc 36:1–7
Mackerras IM, Mackerras MJ, Burnet FM (1940) Experimental studies of ephemeral fever in Australian cattle. Aust Commonw Sci Ind Res Bull 136:1–116
McWilliam SM, Kongsuwan K, Cowley JA, Byrne KA, Walker PJ (1997) Genome organization and transcription strategy in the complex G_{NS}-L intergenic region of bovine ephemeral fever rhabdovirus. J Gen Virol 78:1309–1317
Muller MJ, Standfast HA (1986) Vectors of ephemeral fever group viruses. In: St. George TD, Kay BH, Blok J (eds) Arbovirus research in Australia. Proceedings of the Fourth Symposium. CSIRO/QIMR, Brisbane, pp 295–300
Murphy FA, Taylor WP, Mims CA, Whitfield SG (1972) Bovine ephemeral fever virus in cell culture and mice. Arch Ges Virusforsch 38:234–249
Murray MD (1970) The spread of ephemeral fever of cattle during the 1967–68 epizootic in Australia. Aust Vet J 46:77–82
Nandi S, Negi BS (1999) Bovine ephemeral fever: a review. Comp Immunol Microbiol Infect Dis 22:81–91

Newton LG, Wheatley CH (1970) The occurrence and spread of ephemeral fever of cattle in Queensland. Aust Vet J 46:561–568
Odend'hal S (1983) The geographical distribution of animal viral diseases. Academic Press, New York
Piot JB (1896) Epizootic dengue fever in cattle in Egypt. Prix de l'Académie Nationale de Médecine, France
Riding GA, Wang Y, Walker PJ (1993) A strategy for purification and peptide sequence analysis of bovine ephemeral fever virus structural proteins. In: St. George TD, Uren MF, Young PL, Hoffman D (eds) Bovine ephemeral fever and related rhabdoviruses. ACIAR, Canberra, pp 98–102
St. George TD (1986) The epidemiology of bovine ephemeral fever in Australia and its economic effect. In: St. George TD, Kay BH, Blok J (eds) Arbovirus research in Australia. Proceedings of the Fourth Symposium. CSIRO/QIMR, Brisbane, pp 281–286
St. George TD (1988) Bovine ephemeral fever: a review. Trop Anim Hlth Prod 20:194–202
St. George TD (1990) Bovine ephemeral fever virus. In: Dinter Z, Morein B (eds) Virus infections of vertebrates, Vol 3. Virus infections of ruminants. Elsevier, Amsterdam, pp 405–415
St. George TD (1993) The natural history of ephemeral fever in cattle. In: St. George TD, Uren MF, Young PL, Hoffman D (eds) Bovine ephemeral fever and related rhabdoviruses. ACIAR, Canberra, pp 13–19
St. George TD, Standfast HA (1988) Bovine ephemeral fever. In: Monath TP (ed) The arboviruses: epidemiology and ecology, Vol 2. CRC Press, Boca Raton, pp 71–86
St. George TD, Standfast HA, Christie DG, Knott SG, Morgan IR (1977) The epizootiology of bovine ephemeral fever in Australia and Papua-New Guinea. Aust Vet J 53:17–28
St. George TD, Standfast HA, Cybinski C, Filippich C, Carley JG (1980) Peaton virus: a new Simbu group arbovirus isolated from cattle and *Culicoides brevitarsis* in Australia. Aust J Biol Sci 33:235–243
St. George TD, Cybinski DH, Murphy GM, Dimmock CK (1984) Serological and biochemical factors in bovine ephemeral fever. Aust J Biol Sci 37:341–349
St. George TD, Uren MF, Zakrzewski H (1986) The pathogenesis and treatment of bovine ephemeral fever. In: St. George TD, Kay BH, Blok J (eds) Arbovirus research in Australia. Proceedings of the Fourth Symposium. CSIRO/QIMR, Brisbane, pp 303–307
Schweinfurth G (1878) The heart of Africa, Sampson, Low and Searle, London, pp 280–281
Seddon HR (1938) The spread of ephemeral fever (three-day sickness) in Australia in 1936–37. Aust Vet J 14:90–101
Seddon HR (1966) Ephemeral fever. In: Diseases of domestic animals in Australia. Part 4: protozoan and virus diseases. Commonwealth Department of Health, Canberra, pp 118–126
Shirakawa H, Ishibashi K, Ogawa T (1994) A comparison of the epidemiology of bovine ephemeral fever in South Korea and south-western Japan. Aust Vet J 71:50–52
Snowdon WA (1970) Bovine ephemeral fever: the reaction of cattle to different strains of ephemeral fever virus and the antigenic comparison of two strains of the virus. Aust Vet J 46:258–266
Spradbrow PB (1975) Attenuated vaccines against bovine ephemeral fever. Aust Vet J 51:464–468

Standfast HA, St. George TD, Dyce AL (1976) The isolation of ephemeral fever virus from mosquitoes in Australia. Aust Vet J 52:242

Takematsu M, Sasahara J, Chikatsune M, Okazaki K (1956) Immunological experiments on bovine epizootic fever. Bull Natl Inst Anim Hlth 31:25

Theodoridis A, Boshoff SET, Botha MJ (1973a) Studies on the development of a vaccine against bovine ephemeral fever. Onderstepoort J Vet Res 40:77–82

Theodoridis A, Giesecke WH, Du Toit IJ (1973b) Effects of ephemeral fever on milk production and reproduction of dairy cattle. Ondersteeport J Vet Res 40:83–92

Tian F, Jiang C, Zakrzewski H, Davis SS (1987) A comparison of a Chinese and an Australian strain of bovine ephemeral fever virus. Aust Vet J 64:159

Tomori O, Fagbami A, Kemp G (1974) Kotonkan virus: experimental infection of Fulani calves. Bull Epizoot Dis Afr 22:195–200

Tomori O, Fagbami A, Fabiyi A (1975) Serum antibodies to two rhabdoviruses (bovine ephemeral fever and Kotonkan) in calves on the University of Ibaban agricultural farm. Bull Anim Hlth Prodn Afr 23:39

Tzipori S (1975) The isolation of bovine ephemeral fever virus in cell cultures and evidence for autointerference. Aust J Exp Biol Med Sci 53:273–279

Tzipori S, Spradbrow PB (1973) Studies on vaccines against bovine ephemeral fever. Aust Vet J 49:183–187

Tzipori S, Spradbrow PB (1978) A cell culture vaccine against bovine ephemeral fever. Aust Vet J 54:323–328

Uren MF (1983) Bovine ephemeral fever virus. In: Corner LA, Bagust TJ (eds) Australian standard diagnostic techniques for animal diseases. CSIRO, for Standing Committee on Agriculture and Resource Management, Melbourne, pp 1–8

Uren MF (1989) Bovine ephemeral fever. Aust Vet J 66:233–236

Uren MF, Murphy GM (1992) Studies on the pathogenesis of bovine ephemeral fever in sentinel cattle. II Haematological and biochemical data. Vet Microbiol 10:505–515

Uren MF, St. George TD, Stranger RS (1983) Epidemiology of ephemeral fever in Australia 1975–1981. Aust J Biol Sci 36:91–100

Uren MF, St. George TD, Kirkland PD, Stranger RS, Murray MD (1987) Epidemiology of bovine ephemeral fever in Australia 1981–1985. Aust J Biol Sci 40:125–136

Uren MF, St. George TD, Zakrzewski H (1989) The effect of anti-inflammatory agents on the clinical expression of bovine ephemeral fever. Vet Microbiol 19:99–111

Uren MF, St. George TD, Murphy GM (1992) Studies on the pathogenesis of bovine ephemeral fever in experimental cattle. III Virological and biochemical data. Vet Microbiol 30:297–307

Uren MF, Zakrzewski H, Davis SS (1993) Antibody and cell proliferative responses of cattle vaccinated with bovine ephemeral fever virus proteins. In: St. George TD, Uren MF, Young PL, Hoffman D (eds) Bovine ephemeral fever and related rhabdoviruses. ACIAR, Canberra, pp 122–126

Uren MF, Walker PJ, Zakrzewski H, St. George TD, Byrne KA (1994) Effective vaccination of cattle using the virion G protein of bovine ephemeral fever virus as an antigen. Vaccine 12:845–850

Van der Westhuizen B (1967) Studies on bovine ephemeral fever. I Isolation and preliminary characterization of a virus from natural and experimentally produced cases of bovine ephemeral fever. Onderstepoort J Vet Res 34:29–40

Vanselow BA, Abetz I, Trenfield K (1985) A bovine ephemeral fever vaccine incorporating adjuvant Quil A: a comparative study using adjuvants Quil A, aluminium hydroxide gel and dextran sulphate. Vet Rec 117:37–43

Vanselow BA, Walthall JC, Abetz I (1995) Field trials of ephemeral fever vaccines. Vet Microbiol 46:117–130

Venter GJ, Hamblin C, Paweska JT (2003) Determination of the oral susceptibility of South African livestock-associated biting midges, *Culicoides* species, to bovine ephemeral fever virus. Med Vet Entomol 17:133–137

Walker PJ, Benmansour A, Dietzgen R, Fang R-X, Jackson AO, Kurath G, Leong JC, Nadin-Davies S, Tesh RB, Tordo N (1999) *Rhabdoviridae*. In: van Regenmortel MHV, Fauquet CM, Bishop DHL, Carstens EB, Estes MK, Lemon SM, Miniloff J, Mayo MA, McGeoch DJ, Pringle CR and Wickner RB (eds) Virus taxonomy. Seventh report of the International Committee on Taxonomy of Viruses. Academic Press, New York, pp 563–583

Walker PJ, Kongsuwan K (1999) Deduced structural model for animal rhabdovirus glycoproteins. J Gen Virol 80:1211–1220

Walker PJ, Byrne KA, Cybinski DH, Doolan DL, Wang Y (1991) Proteins of bovine ephemeral fever virus. J Gen Virol 72:67–74

Walker PJ, Byrne KA, Riding GA, Cowley JA, Wang Y, McWilliam S (1992) The genome of bovine ephemeral fever rhabdovirus contains two related glycoprotein genes. Virology 191:49–61

Walker PJ, Wang Y, Cowley JA, McWilliam SM, Prehaud C (1994) Structural and antigenic analysis of the nucleoprotein of bovine ephemeral fever rhabdovirus. J Gen Virol 75:1889–1899

Wang FI, Hsu AM, Huang KJ (2001) Bovine ephemeral fever in Taiwan. J Vet Diagn Invest 13:462–467

Wang YH, Walker PJ (1993) Adelaide River rhabdovirus expresses consecutive glycoprotein genes as polycistronic mRNAs: new evidence of gene duplication as an evolutionary process. Virology 195:719–731

Wang YH, McWilliam SM, Cowley JA, Walker PJ (1994) Complex genome organization in the G_{NS}-L intergenic region of Adelaide River rhabdovirus. Virology 203:63–72

Young PL, Spradbrow PB (1980) The role of neutrophils in bovine ephemeral fever of cattle. J Inf Dis 142:50–55

Zakrzewski H, Cybinski DH (1984) Isolation of Kimberley virus, a rhabdovirus, from *Culicoides brevitarsis*. Aust J Exp Biol Med Sci 62:779–780

Zakrzewski H, Cybinski DH, Walker PJ (1992) A blocking ELISA for the detection of specific antibodies to bovine ephemeral fever virus. J Immunol Meth 151:289–297

CTMI (2005) 292:81–117

Fish Rhabdoviruses: Molecular Epidemiology and Evolution

B. Hoffmann (✉) · M. Beer · H. Schütze · T. C. Mettenleiter

Friedrich-Loeffler-Institut, Federal Research Institute for Animal Health,
17493 Greifswald-Insel Riems, Germany
bernd.hoffmann@fli.bund.de

Abstract Rhabdoviruses may cause serious diseases in wild and farmed fish. Within the *Rhabdoviridae* six genera have been established: *Ephemerovirus, Cytorhabdovirus, Nucleorhabdovirus, Lyssavirus, Vesiculovirus,* and *Novirhabdovirus.* Viruses that infect fish are official or tentative members of the genera *Vesiculovirus* and *Novirhabdovirus*, or are listed as unassigned rhabdoviruses. In this report, we summarize and discuss published and our own unpublished data on the molecular epidemiology and phylogeography of fish rhabdoviruses including intrapopulational differences and subgrouping of fish rhabdoviruses, in particular the species *spring viremia of carp virus* (SVCV), *infectious hematopoietic necrosis virus* (IHNV) and *viral hemorrhagic septicemia virus* (VHSV).

1 Introduction

Modern studies of viral phylogenetics and evolution rely almost completely on data derived from sequencing of viral genomes with PCR-based methods offering simplicity, sensitivity, specificity and speed. In addition to PCR-based approaches, which generally focus on limited areas of a genome and are particularly powerful in molecular epidemiology, sequence information on whole viral genomes has been continuously accumulating over the past decades. This wider fundament of information ensures that our understanding of viral phylogeny and evolution is soundly based. It is now a relatively straightforward route between discovery of a new virus and derivation of its complete sequence, regardless of genome size. Moreover, there are numerous examples of sequence information for multiple strains of the same viral species. Except for the essential information on individual viruses, current understanding of gene content, phylogeny and evolution is built upon extensive sequence information obtained from related organisms. The result of increased sequencing activities is an extensive knowledge of molecular relationships, now available for most virus families in the form of phylogenetic trees. By illustrating the relationship between viruses, these trees provide us with an understanding of past and ongoing viral evolution.

For fish rhabdoviruses the amount of available sequence information has increased significantly in the past several years. By mid 2003 a total of ten complete genome sequences of fish rhabdoviruses have been submitted to GenBank, in addition to a large number of partial sequences. The fish rhabdoviruses are a group of RNA viruses that is widely distributed in marine and freshwater fish all over the world. Members of this group share a number of distinct features, including a negative-strand RNA genome encoding at least five open reading frames in the order nucleoprotein (N), phosphoprotein (P), matrixprotein (M), glycoprotein (G) and RNA polymerase (L) (reviewed in Bernard and Bremont 1995). Whereas vesiculoviruses contain only these five classical genes, members of the genus *Novirhabdovirus* are characterized by an additional gene that is located between the glycoprotein and RNA polymerase genes. This gene encodes a nonstructural, nonvirion protein (NV) and is unique to this genus.

Phylogenetic analysis has been an increasingly important tool in the investigation of the viral epidemiology by its ability to identify speciation, geographical links and common ancestry. Here, we performed a phylogenetic analysis of the fish rhabdoviruses within the *Rhabdoviridae* family based on new as well as previously published sequence information. Moreover, evolutional and phylogeographical aspects of the fish rhabdoviruses will be discussed.

2
Taxonomic Grouping of Fish Rhabdoviruses Within the *Rhabdoviridae* Family

Historically, fish rhabdoviruses were first assigned to either the genus *Vesiculovirus* or the genus *Lyssavirus*, based on their protein profiles after gel electrophoresis. These genera contain many important mammalian rhabdoviruses such as *rabies virus*and *vesicular stomatitis virus* . Until the late 1980s these were the only two genera within the *Rhabdoviridae* family, which were considered to encompass all rhabdoviruses.

However, as more viral genomes became characterized by sequence analysis, several new genera within the *Rhabdoviridae* family have been established. The sixth report of the International Committee for Virus Taxonomy (ICTV) lists the three additional genera *Ephemerovirus*, *Cytorhabdovirus* and *Nucleorhabdovirus* (Wunner et al. 1995). By the seventh report of the ICTV, the new *Novirhabdovirus* genus was established (Walker et al. 2000). The present six accepted genera within the *Rhabdoviridae* are *Vesiculovirus, Lyssavirus, Ephemerovirus, Cytorhabdovirus, Nucleorhabdovirus,* and *Novirhabdovirus.* In Table 1 the six genera including important members of each genus are listed with fish rhabdoviruses depicted in italics.

Molecular phylogenies determined by using N, G or L gene sequences support the integrity of the *Rhabdoviridae* family, and the classification of species within the established genera. However, low sequence identities of the G gene prevent the construction of a general phylogenic tree, whereas sequences of the L gene, which is highly conserved, are at present not available for the type species of each genus. Therefore, a universal phylogenetic tree of the *Rhabdoviridae* can best be constructed by using sequences of the reasonably conserved N protein gene. Direct comparison of *Lyssavirus* genes demonstrates that the degree of conservation decreases in the order nucleoprotein >matrix protein >glycoprotein >phosphoprotein (Bourhy et al. 1993).

Therefore, available GenBank sequence data of the N gene were used for a taxonomic grouping of the fish rhabdoviruses within the family *Rhabdoviridae.* The abbreviation of the analyzed rhabdoviruses and the accession numbers of the N gene sequence data are described in Table 1. In most cases, complete N gene sequences were submitted, but for several viruses (ABLV, DUVV, EBLV-1, EBLV-2, LBV, MOKV, MFSV, WRSV) only partial sequence data of the N gene were available. Therefore, a minimum of 1,337 nucleotides of the N genes were used to create the phylogenetic tree (Fig. 1).

Sequence data were assembled and analyzed using HUSAR (Heidelberg Unix Sequence Analysis Resources, German Cancer Research Center [DKFZ],

Table 1 Taxonomic classification of the *Rhabdoviridae* family

Genus	Host	Important members of the genera and assignment of the fish rhabdoviruses[a]	Abbreviation	Accession number[c]
Vesiculovirus	Mammals	Vesicular stomatitis Indiana virus (TS)	VSIV	NC_001560
		Vesicular stomatitis New Jersey virus	VSNJV	M31846
	Fish	*Spring viremia of carp virus*[b]	SVCV	AJ318079
		Eel virus American[b]	EVA	
		Eel virus European X[b]	EVEX	
		Pike fry rhabdovirus[b]	PFRV	
		Ulcerative disease rhabdovirus[b]	UDRV	
Lyssavirus	Mammals	Rabies virus (TS)	RABV	NC_001542
		Mokola virus	MOKV	U22843
		Lagos bat virus	LBV	U22842
		European bat lyssavirus 1	EBLV-1	U22844
		European bat lyssavirus 2	EBLV-2	U22846
		Duvenhage virus	DUVV	U22848
		Australian bat lyssavirus	ABLV	NC_003243
Ephemerovirus	Mammals	Bovine ephemeral fever virus (TS)	BEFV	NC_002526
		Adelaide River virus	ARV	U10363
Novirhabdovirus	Fish	*Infectious hematopoietic necrosis virus (TS)*	IHNV	X89213
		Viral hemorrhagic septicemia virus	VHSV	NC_000855
		Hirame rhabdovirus	HIRRV	AF104985
		Snakehead rhabdovirus[b]	SHRV	NC_000903
		Eel virus B12[b]	EV-B12	
		Eel virus C26[b]	EV-C26	

Table 1 (continued)

Genus	Host	Important members of the genera and assignment of the fish rhabdoviruses[a]	Abbreviation	Accession number[c]
Cytorhabdovirus	Plants	Lettuce necrotic yellows virus (TS)	LNYV	L30103
		Northern cereal mosaic virus	NCMV	NC_002251
Nucleorhabdovirus	Plants	Potato yellow dwarf virus (TS)	PYDV	
		Rice yellow stunt virus	RYSV	NC_003746
		Sonchus yellow net virus	SYNV	NC_001615
		Maize fine streak virus[b]	MFSV	AF518002
Unassigned species	Animals	Flanders virus	FLAV	AF523194
		Sigma virus	SIGMAV	X91062
	Fish	*Sea trout rhabdovirus 28/97*	STRV-28/97	AF434992
		Trout rhabdovirus 903/87	TRV-903/87	AF434991
	Plants	Wheat rosette stunt virus	WRSV	AF059603

TS, Type species.
[a]Fish rhabdoviruses are depicted in italics.
[b]Tentative member of the genus.
[c]Sequence data from these accession numbers were used to construct the phylogenetic tree of the *Rhabdoviridae* family depicted in Fig. 1.

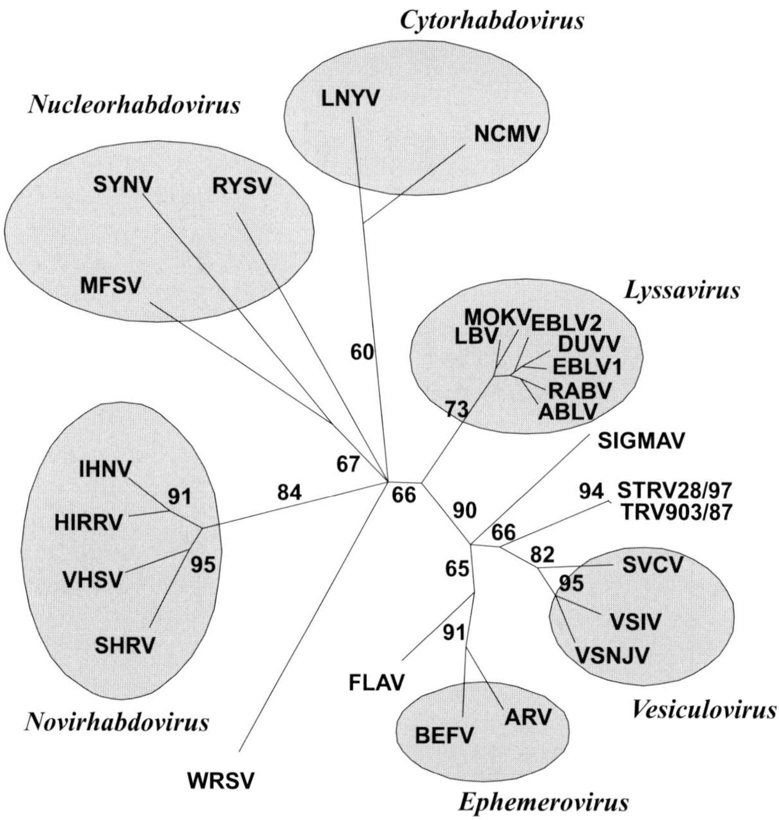

Fig. 1 Phylogenetic tree based on the N gene sequences of rhabdoviruses using the software PUZZLE (Strimmer and von Haeseler 1996) and TREEVIEW (Page 1996). The *abbreviations* of the viruses and accession number of used sequence data are listed in Table 1. The *numbers* refer to the support for the internal branches of the quartet puzzling tree in percent (1,000 puzzling steps). Branch lengths are proportional to genetic distances

Heidelberg). Phylogenetic analyses of the aligned data (ClustalW) were performed using the software PUZZLE (Strimmer and von Haeseler 1996, 1997).

For maximum likelihood tree reconstructions, 1,000 puzzling steps and the Hasegawa model for nucleotide substitution were used. The phylogenetic tree generated by PUZZLE was visualized using the software TREEVIEW (Page 1996). A comparison of sequence identities of the N genes of rhabdoviruses was obtained with the program BESTFIT (HUSAR, DKFZ).

Our phylogenetic analyses confirmed the established classification of rhabdoviruses into six genera. The *Nucleorhabdovirus* and *Cytorhabdovirus* genera contain plant pathogenic viruses (SYNV, RYSV, LNYV, NCMV) for which the classification is primarily based on the site of virus maturation (cytoplasm: *Cytorhabdovirus*; nucleus: *Nucleorhabdovirus*). Comparative sequence analyses of the plant rhabdoviruses showed no significant identities (Wetzel et al. 1994; Choi et al. 1992; Heaton et al. 1989). However, our phylogenetic analyses (Fig. 1) are impaired by low sequence homology reflected by low bootstrap values. Bootstrap estimates of 70% or less correspond to less than 95% probability that a particular clade exists (Hillis and Bull 1993). Therefore, this assignment may have to be modified in the future and for plant rhabdoviruses further studies on the molecular level are necessary to confirm or to revise the current grouping.

In contrast, results for the genera *Lyssavirus* and *Ephemerovirus* are clear. Both genera contain only rhabdoviruses with mammalian hosts, and the phylogenetic tree depicted in Fig. 1 indicates a distinct grouping of members of the genus *Lyssavirus* and members of the genus *Ephemerovirus* with significant probabilities values that provide a high level of confidence into the phylogenetic analysis.

Some of the most important fish viruses are members of the *Vesiculovirus* and *Novirhabdovirus* genera, as confirmed by our phylogenetic analysis based on the complete N gene sequences (Fig. 1). In particular, members of the *Novirhabdovirus* genus formed a distinct clade, which was well supported by bootstrap analysis. The aquatic vesiculo-type viruses are clustered with or near the *Vesiculovirus* genus. A more detailed phylogenetic and phylogeographic analysis of selected fish rhabdoviruses follows below.

3
Fish Rhabdoviruses Within the *Vesiculovirus* Genus

Vesicular stomatitis Indiana virus (VSIV) is the type strain of the genus *Vesiculovirus*, which includes eight addition viral species that infect mammals. Tentative members of the genus are the fish viruses *spring viremia of carp virus* (SVCV), *pike fry rhabdovirus* (PFRV), *ulcerative disease rhabdovirus* (UDRV), *eel virus Europe X* (EVEX) and *eel virus American* (EVA) (Walker et al. 2000). Recently, partial genomic sequences of two new fish rhabdoviruses have been published (Johansson et al. 2002, 2001). This *trout rhabdovirus* 903/87 (TRV-903/87) and the closely related *sea trout rhabdovirus* 28/97 (STRV-28/97) showed the highest sequence homology to vesiculoviruses.

3.1 Spring Viremia of Carp Virus

The most important fish virus of this genus is Spring viremia of carp virus SVCV or *Rhabdovirus carpio* (Fijan et al. 1971). SVCV causes a severe hemorrhagic disease of cyprinids, which is notifiable to the Office International des Epizooties (OIE). Recently, Ahne et al. (2002) published a comprehensive review on SVC and the antigenic properties and genetic structure of its causative agent.

Characterization of the genomic RNA of SVCV started in 1984 with the publication of sequences of the M gene and the first 70 nucleotides at the 3′ end of the genome (Kiuchi and Roy 1984; Roy et al. 1984). More intensive investigations on SVCV yielded partial sequences of the L gene, the complete G gene and the gene junctions of SVCV (Bjorklund et al. 1995, 1996). In 1999, Johnson et al. published the complete nucleotide sequence of the glycoprotein gene of a rhabdovirus isolated from diseased shrimps in Hawaii (Lu et al. 1991). Sequence comparison of the G protein nucleotide sequence of this shrimp isolate reached 99% identity to the corresponding sequence of SVCV. Thus, the so-called rhabdovirus of penaeid shrimp (RPS) is actually an SVCV strain, and the identification of RPS was the first report of SVCV isolation outside of Europe and Asia. It was also the first report of a SVCV infecting a non-vertebrate species.

In 2002, two complete sequences of the SVCV genome were reported (Hoffmann et al. 2002 – accession number AJ318079; Bjorklund et al. – accession number U18101). The SVCV genome contains a 59-base putative leader region at the 3′ terminus followed by the universal consensus start signal (AACAG, mRNA-sense). This short nucleotide sequence is used for the transcription start of all SVCV genes. As is true for other members of the genus, the SVCV virion contains the five proteins N, P, M, G and L. The genome is further characterized by conserved gene junctions and short intergenic regions of two to four nucleotides. The transcription stop/polyadenylation signal of the SVCV genes (TATG(A)7) is strictly conserved (Bjorklund et al. 1996) and the SVCV L gene is followed by a 46-base trailer that forms the 5′ terminus of the genome. A gene encoding a non-virion (NV) protein between the G and L gene is not present in the SVCV genome (Walker et al. 2000; Kurath et al. 1995, 1997; Schütze et al. 1996). Similarly, it does not specify a large non-coding region between the G and the L genes, as found in lyssavirus genomes (Tordo et al. 1986). Thus, the overall genome structure and the gene order of SVCV matches that of the *Vesiculovirus* genus.

Ahne et al. (2002) compared the N and G gene sequences of SVCV with corresponding sequences of 15 other rhabdoviruses. The deduced phylogenetic trees confirmed the assignment of SVCV to the genus *Vesiculovirus*. The highest amino acid identity of the deduced G proteins of 31.2%–33.2% was found with *vesicular stomatitis Indiana virus* (VSIV), *vesicular stomatitis New Jersey virus* (VSNJV) and *chandipura virus* (CHPV), which are established members of the *Vesiculovirus* genus. Despite minor differences in the topology of the phylogenetic trees generated with sequences of the N and the G genes, both trees indicate a close phylogenetic relationship between SVCV and the accepted members of the *Vesiculovirus* genus. This is further supported by significant bootstrap values, providing a high level of confidence in this relationship. A similar phylogenetic analysis of rhabdoviral G proteins was presented by Johnson et al. (1999).

A more detailed phylogenetic analysis of several SVCV isolates was performed by Stone et al. (2003), who investigated the homology between SVCV and PFRV. For this study, a 550-nt region of the glycoprotein gene from 36 putative SVCV and PFRV isolates from different geographical locations and fish species was sequenced. The phylogenetic tree generated by maximum parsimony analysis based on these partial G gene sequences identified four genogroups that were supported by high bootstrap values (>97%) for all branches. Interestingly, no correlation was observed between the genogroups and the viral host species. However, all 15 isolates previously identified as SVCV based on serological experiments were assigned to genogroup I despite a higher-than-expected diversity at the nucleotide level of 82.7%–100% (Stone et al. 2003). The authors interpreted these results as evidence for a probable independent evolution of SVCV in several different geographical areas. Phylogenetic re-analyses of the G gene sequence data of the investigated 15 SVCV by maximum parsimony and neighbor-joining methods identified four different subgroups (Ia-d) supported by bootstrap values of >98% (Stone et al. 2003). Asian SVCV isolates fell into subgroup Ia, isolates from Moldavia, the Ukraine and Russia were assigned to subgroups Ib and Ic, and those from the UK to subgroup Id. Because several isolates from areas within the former USSR also clustered with the European subgroup Id, further investigations of different SVCV isolates from all over the world are necessary to confirm and complete the sub-grouping of SVCV isolates. For a correct depiction of the phylogenetic and geographical relationship of strains within the species SVCV, other SVCV genes should also be analyzed.

We sequenced the complete phosphoprotein genes of six wild-type strains of SVCV isolated in the years 2000 and 2001 in Germany. Five isolates originated from koi carp (*Cyprinus carpio koi*) and one came from a common carp (*Cyprinus carpio carpio*). The complete P genes including adjacent gene junctions were amplified by RT-PCR. After cloning of the PCR fragments generated by least two independent RT-PCR assays, sequences were determined using cycle sequencing of at least six different clones. The GenBank accession numbers of the P genes of the SVCV isolates 01/01-V1592, 01/01-V1621, 19/00-52/94, 19/00-59/95, 19/00-73/94 and 17/00-47/3 reported in this study are AY424883–AY424888. The first morphological and serological characterizations of the isolates 01/01-V1592 and 01/01-V1621 were published by Neukirch and Kunz (2001). Comparison of the phosphoprotein gene sequences of the published full length SVCV genome (Hoffmann et al. 2002, accession number AJ318079; Bjorklund et al., accession number U18101) with the P gene sequences of these new isolates confirmed the same length as the published P gene sequences, and sequence identity ranging between 98.4% and 100%. The highest divergence of the P gene sequences with 17 nucleotide exchanges was observed between the published data of Bjorklund et al. and the isolate 01/01-V1621 from koi carp (accession number AY424884), whereas the P gene nucleotide sequences of the koi carp isolates 19/00-52/94 (accession number AY424885) and 19/00-73/94 (accession number AY424887) are identical. These two SVCV strains were isolated during different outbreaks in the same geographical area in 2000.

The amino acid sequences of the P proteins are also most highly conserved within the field isolates. The highest number of amino acid exchanges (7 of 309 amino acids) was observed between the P protein of an SVCV strain published by Hoffmann et al. (2002, accession number AJ318079) and the P protein of isolate 01/01-V1621 from a koi carp, and thus the amino acid sequences were overall still 97.7% identical.

Based on the results of Bourhy et al. (1993) for the *Lyssavirus* genus, a low degree of conservation for the P gene compared to the other viral proteins was expected, which should help in assessing intrapopulational divergence of SVCV isolated in a restricted area. The lowest homology (5 divergent of 309 amino acids=98.4%) was observed between the deduced P protein of isolate 01/01-V1621 from the koi carp and the P protein of isolate 17/00-47/3 from diseased common carp. However, no correlation between amino acid exchanges and the host species could be deduced, because the P protein sequences of other koi carp isolates also differ in only one to three amino acids from the P protein of the common carp isolate 17/00-47/3. This confirms data from the analysis of parts of the G gene of SVCV isolates from common carp, koi carp, silver carp, bighead carp, grass carp and from rainbow trout, in which also no correlation between genetic grouping and the host species was found (Stone et al. 2003).

An interesting observation in both available complete SVCV sequences is a nucleotide substitution (C to G) at position −4 from the P gene start codon. In all six analyzed field isolates, the consensus sequence of the N–P gene junction was TATG(A)$_7$CTAACAGA**G**ATC, where TATG(A)$_7$ represents the polyadenylation signal of the N gene and AACAG the transcription initiation sequence of the P gene. Thus, in all investigated wild-type isolates of SVCV, a G was present at the −4 position of the P protein start codon, which confirms this unique feature compared to the −4 position of the other SVCV genes. Another interesting feature could be observed in the P–M gene junctions of the six isolates. The transcription start signal of the three isolates 19/00-52/94, 19/00-59/95, and 17/00-47/3 was CACAG instead of the commonly known transcription start signal AACAG of the *Vesiculovirus* genus (Rose 1980; Rose and Gallione 1981). Whether transcription of the M gene is affected by this substitution has to be determined.

In summary, analysis of SVCV isolates based on glycoprotein gene sequences resulted in the establishment of four different subgroups (Stone et al. 2003). These subgroups, Ia to Id, correlate in the majority of cases with the area of isolation. Nevertheless, analyses of more isolates from all over the world are necessary for a confirmation of the described SVCV subgrouping.

3.2
Pike Fry Rhabdovirus

Pike fry rhabdovirus (PFRV) was isolated from moribund *Esox lucius* L. during an outbreak of red disease, an acute condition characterized by hemorrhagic lesions on the trunk, ascites and high rates of mortality, in a Dutch pike hatchery in 1973 (de Kinkelin et al. 1973). In the seventh report of the ICTV (Walker et al. 2000), PFRV was also designated as grass carp rhabdovirus because an antigenically related virus was isolated from grass carp (*Ctenopharyngodon idella)* (Ahne 1975). However, PFRV was also obtained from other fish species (Rowley et al. 2001; Dixon et al. 1994; Haenen and Davidse 1989; Ahne and Thomson 1986; Fijan et al. 1984, Ahne et al. 1982).

The first data on different PFRV strains isolated in Northern Ireland, Republic of Ireland, The Netherlands and France were presented by Rowley et al. (2001) based on sequences of nucleotides 405–954 of the glycoprotein gene of seven isolates. Sequence alignments revealed a high degree of identity (99.3%–100%) at the nucleotide level between the isolates from Northern Ireland and the Republic of Ireland. A high degree of identity was also found between these five isolates and an isolate from The Netherlands (96.6%–97.1%), whereas the PFRV reference strain F4 from France exhibited less than 82.4% sequence homology to the other strains. The sequence of the investigated SVCV reference strain S30 had an identity of only 67.7% when compared to the PFRV isolates from Northern Ireland and the Republic of Ireland (Rowley et al. 2001). Based on these data, the authors identified three distinct genogroups of fish viruses within the *Vesiculovirus* genus. The reference strains of SVCV and PFRV were the sole members of two of these genogroups, and the six isolates from Northern Ireland, the Republic of Ireland and The Netherlands clustered into the third.

A more comprehensive investigation of different PFRV isolates was performed by Stone et al. (2003), which confirmed the previous study of Rowley et al. (2001). The phylogenetic analysis of 21 PFRV isolates using a 550-bp fragment of the glycoprotein gene showed presence of three genogroups (II–IV). The majority of the PFRV isolates (19 of 21) were assigned to genogroup IV. These isolates showed a high degree of nucleotide sequence identity to each other (>93.7%). Genogroup III was formed by the PFRV reference strain F4, which exhibited a homology of less than 80% to genogroup IV strains. This was considered to be sufficiently different for establishing a separate group. A grass carp isolate V76, which had also previously been identified as PFRV (Ahne 1975), shared less than 70% nucleotide sequence identity with both PFRV F4 (genogroup III) and representatives of genogroup IV, and was therefore assigned to genogroup II. Comparison of the amino acid sequences yielded the highest homology between this grass carp isolate V76 and the SVCV reference strain S30 (70.3%). In contrast, the PFRV reference strain F4 (genogroup III) and a representative of the genogroup IV (PFRV-S64) showed only 64.6% and 56.3% amino acid sequence identity, respectively. From these studies, it is possible that the grass carp isolate V76 (genogroup II) constitutes an evolutionary link between SVCV (genogroup I) and PFRV (genogroups III, IV).

3.3 Other Aquatic Vesiculo-Type Viruses

Epizootic ulcerative syndrome (EUS) is a seasonal epizootic disease of great importance in wild and farmed fresh- and brackish-water fish. The disease has a complex infectious etiology but the most probable causative agents of EUS are the fungi *Aphanomyces (A.) invadans* and *A. piscicida*. The main clinical signs are necrotic ulcerative lesions of muscle tissue, but in some cases fungal hyphae extend into the visceral organs. Despite the fungal etiology, co-factors have been postulated as necessary for initiation of the disease including epidermal damage, environmental stress, and bacterial and viral infections (Lilley et al. 2003; Mastan and Qurehi 2001; John et al. 2001; Mukherjee et al. 1995; Lilley and Frerichs 1994). In 1986, different authors (Frerichs et al.; Wattanavijarn et al.) described the isolation of rhabdoviruses from fish exhibiting EUS in South-east Asia. Furthermore, the detection of *ulcerative disease rhabdovirus* (UDRV) and other rhabdoviruses (Lio-Po et al. 2000; Ahne et al. 1988) as well as birnaviruses and reoviruses were described at an outbreak on EUS (Frerichs 1995). The isolated rhabdoviruses were subsequently grouped into two different serotypes (Kasornchandra et al. 1992). The northern Thailand strains (Frerichs et al. 1986) designated as UDRV, and the isolates from central Thailand (Wattanavijarn et al. 1986) were defined as *snakehead rhabdovirus* (SHRV). Furthermore, on the basis of structural polypeptide mobility patterns, it was suggested that the UDRV isolates exhibited a *Vesiculovirus*-type protein profile (Kasornchandra et al. 1992). Since no sequence data of the UDRV genome are available, a specific taxonomic assignment based on the nucleotide sequence is not possible.

The *eel virus European X* (EVEX) and *eel virus American* (EVA) are also fish pathogens. To date they are assigned to the vesiculo-type viruses based on morphological and serological data (Sano et al. 1977a; Sano 1976). Sequences of the viral genomes are not available yet.

Recently, two new fish rhabdoviruses have been identified. *Trout rhabdovirus* 903/87 (TRV-903/87) was isolated from moribund brown trout (*Salmo trutta lacustris* L.) fingerlings in Northern Finland in 1987 (Koski et al. 1992). The electrophoretic migration pattern of the five viral structural proteins suggested that the virus is similar to the vesiculoviruses. TRV-903/87 is serologically related to perch rhabdovirus (Dorson et al. 1984) and pike rhabdovirus DK 5533 (Jorgensen et al. 1993), but exhibited a different protein migration pattern. It was concluded that TRV-903/87 constitutes a novel pathogenic fish rhabdovirus (Björklund et al. 1994; Koski et al. 1992). The second new rhabdovirus, preliminarily designated as *sea trout rhabdovirus* 28/97 (STRV-28/97), was isolated from diseased sea trout (*Salmo trutta trutta*) from the archipelago of Stockholm, Sweden, in 1996. STRV-28/97 was found to be serologically related to TRV-903/87 (Björklund et al. 1994). Sequence data of the N, P, M, G and L genes of both virus genomes (Johansson et al. 2001, 2002) showed that the gene order, the length of single genes and the putative transcription start and stop signals were similar to that of SVCV and VSIV. The obtained partial sequence of the gene following the glycoprotein encoding sequences closely resembles viral polymerases of the other rhabdoviruses, which indicates that both TRV-903/87 and STRV-28/97 lack a NV gene. Thus, the new virus isolates TRV-903/87 and STRV-28/97 are clearly not members of the new *Novirhabdovirus* genus. This has also been confirmed by comparison of the nucleotide and amino acid sequences of all viral proteins of TRV-903/87 and STRV-28/97 with other members of the *Rhabdoviridae*. The overall similarity of the TRV-903/87 and STRV-28/97 proteins is 97.6%–100%. Comparison of the nucleoprotein and glycoprotein sequences to other rhabdoviruses yielded the highest homology with members of the *Vesiculovirus* genus. The nucleoprotein genes were 43%–44% homologous, similar to those of vesiculoviruses, whereas homology to the novirhabdoviruses was only between 26% and 32%. Comparison of the glycoprotein sequences showed a similarity of 35%–40% with vesiculoviruses and of 30%–35% with novirhabdoviruses. However, comparative analysis of the phosphoprotein and matrix protein sequences yielded different results. The phosphoproteins of TRV-903/87 and STRV-28/97 were 29%–36% similar to vesiculoviruses and 32%–37% similar to the novirhabdoviruses. In addition, the matrix protein sequences showed a similarity to vesiculovirus sequences (30%–34%), which was lower than that to novirhabdoviruses (31%–38%).

Phylogenetic studies based on protein parsimony analysis indicate that the nucleoprotein, the glycoprotein and the first part of the polymerase protein (267aa) of strains TRV-903/87 and STRV-28/97 clustered within the *Vesiculovirus* genus. The phosphoproteins of both viruses clustered next to vesiculoviruses, whereas the matrix proteins formed a distinct group. In nucleotide parsimony analysis, identical results were obtained for the N and G genes, whereas the phosphoprotein gene and the matrix protein gene formed distinct groups. Very similar results were obtained in the neighbor distance analysis (Johansson et al. 2002).

In summary, TRV-903/87 and STRV-28/97 are genetically closely related and the gene order for both viruses is N-P-M-G-L. Using phylogenetic analyses, it could be demonstrated that both viruses are most likely grouped within the *Vesiculovirus* genus, although the overall similarity of TRV-903/87 and STRV-28/97 to other vesiculoviruses is not very high, in particular in the phospho and the matrix protein.

3.4
General Remarks for the Aquatic Vesiculo-Type Viruses

In summary, the G and N gene sequences of SVCV, PFRV, TRV-903/87 and STRV-28/97 cluster within the mammalian vesiculoviruses (Fig. 1; Ahne et al. 2002; Johansson et al. 2001, 2002; Rowley et al. 2001; Johnson et al. 1999; Björklund et al. 1996), indicating a relationship of these fish pathogens to the *Vesiculovirus* genus. However, comparison of the equivalent region of the glycoprotein from a number of rhabdoviruses, including vesiculoviruses infecting terrestrial animals, the same high degree of amino acid sequence identity was not observed (Stone et al. 2003), and, based on the comparison of sequences of the P and M genes, the grouping of the these fish viruses within the *Vesiculovirus* genus may not yet be definite. Stone et al. (2003) refer to a comprehensive sequencing program to establish if the aquatic vesiculo-type viruses, including those described by others (Dorson et al. 1984; Johansson et al. 2001), exhibit the relevant properties to be formally recognized as species within the *Vesiculovirus* genus, or whether they demonstrate sufficiently distinct properties to place them into a separate genus within the *Rhabdoviridae* family.

For our evaluation of the taxonomy of the vesiculo-type viruses, the phosphoprotein genes of vesicular stomatitis Indiana virus (VSIV), vesicular stomatitis New Jersey virus (VSNJV), Piry virus (PIRYV), Flanders virus (FLAV), Sigma virus (SIGMAV), TRV-903/87, STRV-28/97 and two different isolates of SVCV were investigated (Fig. 2). As shown in the phylogenetic analysis of the

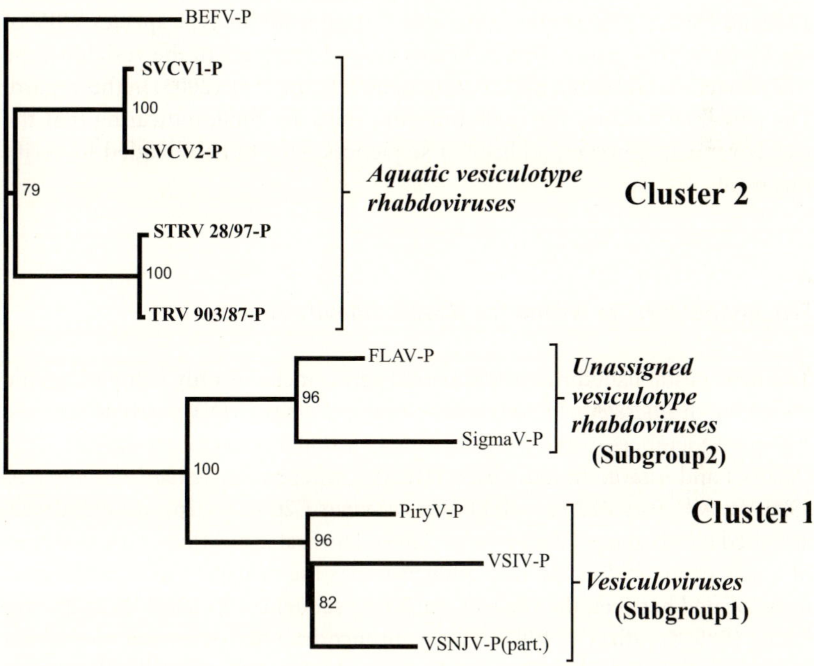

Fig. 2 Phylogenetic tree based on the P gene sequences of rhabdoviruses using the software PUZZLE (Strimmer and von Haeseler 1996) and TREEVIEW (Page 1996). The *abbreviations* of the viruses used are listed in Table 1. For construction of the phylogenetic tree, the following sequence data of the P gene were used (Accession numbers: BEFV – AF234533, SVCV1 – U18101, SVCV2 – AY424884, STRV-28/97 – AF434992, TRV-903-87 – AF434991, FLAV – AF523195, SIGMAV – X91062, PIRYV – D26175, VSIV – J02428, VSNJV – AF252253). The *numbers* refer to the support for the internal branches of the quartet puzzling tree in percent (1,000 puzzling steps). Branch lengths are proportional to genetic distances. BEFV was used as an outgroup root

N genes of rhabdoviruses described above, the PUZZLE software was used (Strimmer and von Haeseler 1996, 1997), and visualized a tree using the software TREEVIEW (Page 1996). For the phylogenetic analysis of the P genes, SVCV sequences published by Bjorklund et al. (accession number U18101) and of the isolate 01/01-V1621 (accession number AY424884) were used, since the P genes of both strains were the most divergent of all published P genes of SVCV. The P gene of bovine ephemeral fever virus was used as an outgroup.

As a result, the phosphoprotein genes of the vesiculo-type viruses represent two distinct clusters (Fig. 2). The first cluster was generated by two subgroups. One subgroup is formed by the official members of the *Vesiculovirus* genus (VSIV, VSNJV, PIRYV) and the other contains two unassigned species (FLAV, SIGMAV). The fish rhabdoviruses SVCV, TRV-903/87 and STRV-28/97 formed a second cluster, phylogenetically more distant from the type species VSIV of the *Vesiculovirus* genus. This indicates a novel grouping of the vesiculo-type rhabdoviruses and confirms the statement of Stone et al. (2003) in this regard. The percentage values for 1,000 puzzling steps are high, indicating that the tree is robust. However, additional sequences have to be included to verify this result.

4
Fish Rhabdoviruses Within the *Novirhabdovirus* Genus

The newly established *Novirhabdovirus* genus includes only fish pathogenic viruses with *infectious hematopoietic necrosis virus* (IHNV) as type species. Accepted members of the genus are also *viral hemorrhagic septicemia virus* (VHSV) and *hirame rhabdovirus* (HIRRV), whereas *snakehead rhabdovirus* (SHRV), *eel virus* B12 (EV-B12) and *eel virus* C26 (EV-C26) are tentatively assigned to this genus (Walker et al. 2000). The first indication of the presence of a new genus within the rhabdoviruses was the discovery of an additional gene located between the G and L encoding sequences in IHNV (Kurath and Leong 1985; Kurath et al. 1985). This gene encodes a nonstructural non-virion protein, designated NV, whose product was identified in infected cells but was absent from purified virions (Schütze et al. 1996).

In the Sixth report of the International Committee on Taxonomy of Viruses (Wunner et al. 1995) HIRRV, IHNV and VHSV are listed as unassigned rhabdoviruses. Results from different laboratories highlighted the only distant relationship between these fish rhabdoviruses and the mammalian rhabdoviruses (Morzunov et al. 1995; Wang et al. 1995; Wang and Walker 1993; Benmansour et al. 1994), and Björklund et al. (1996) suggested the name *Aquarhabdovirus* for a new genus incorporating these viruses, to reflect their aquatic origin. In the Seventh report of the International Committee on Taxonomy of Viruses (Walker et al. 2000), the genus name *Novirhabdovirus* was assigned based on the presence of the NV gene. Novi- is comprised of no- and vi-, which are sigla for non-virion. Therefore, all rhabdoviruses in this genus have an NV gene located between the G and L genes. In addition, species within the *Novirhabdovirus* genus contain nucleotide sequences that cluster phylogenetically with all identified members of the genus, but are more different from rhabdoviruses of other genera. To date, all members of the *Novirhabdovirus* genus are fish pathogens. The function of the NV protein is not known, but the conservation of an open reading frame in diverse virus species and strains may be indicative of a significant biological role.

4.1 Infectious Hematopoietic Necrosis Virus

Infectious hematopoietic necrosis virus (IHNV) is the causative agent of an acute systemic disease inflicting high mortality, especially in young cultured and wild salmonid fish. The virus was first discovered in sockeye salmon (*Oncorhynchus nerka*) dying at hatcheries in Washington in 1953 (Rucker et al. 1953). Similar outbreaks among hatchery-reared Chinook salmon in California were reported in the following decades (Grischkowsky and Amend 1976; Wingfield et al. 1969; Guenther et al 1959). Originally, IHNV was thought to be confined to salmonid fish in the Pacific Northwest of the United States (McAllister 1979). However, the virus spread in the 1970s to Europe, Japan, the eastern US, Korea, Taiwan and China by the shipment of infected fish and contaminated eggs (Bovo et al. 1987; Laurencin 1987; Sano et al. 1977b).

First physicochemical and serological analyses of IHNV were performed by Hill (1975). To date, the biology, immunology and molecular biology of IHNV is well characterized (for reviews, see Bootland and Leong 1999; Wolf 1988). Leong et al. (1981) investigated different IHNV strains from California, Oregon, Washington and Alaska and identified differences between the isolates based on protein profiles. In the 1980s, six viral mRNAs were identified, in contrast to five mRNAs found in other rhabdovirus infected cells (Kurath and Leong 1985; Kurath et al. 1985). The sixth IHNV mRNA was subsequently shown to enocode a unique nonstructural protein that is synthesized in infected cells but excluded from mature virions (Schütze et al. 1996; Kurath and Leong 1985, 1987; Kurath et al. 1985). A more comprehensive comparison of NV nucleotide and amino acid sequences was performed by Kurath et al. (1995, 1997). The authors confirmed the generally low homology of NV sequences compared to the other viral proteins of IHNV, HIRRV and VHSV.

For instance, the identity of the G proteins of IHNV and HIRRV was 74.3%, whereas the identity of the NV protein amounted to only 54.1%. Comparison of the G and NV proteins of IHNV and VHSV showed an identity of 38.8% and 23.3%, respectively. This suggests a relatively low evolutionary constraint on the NV gene.

To date two complete genomic sequences of IHNV are available (Schütze et al. 1995 – accession number X89213; Morzunov et al. 1995 – accession number L40883) and more than 50 partial sequences were submitted to GenBank. Using deduced M, G and L proteins of IHNV and other rhabdoviruses, but also other non-segmented negative-strand RNA viruses (*Mononegavirales*), Morzunov et al. (1995) investigated their phylogenetic relationship. Due to the relatively large genetic distances between different members of the *Mononegavirales* order, phylogenetic analysis of viruses of this broad group was based on amino acid differences found between the relatively conserved L proteins. IHNV was found within a cluster containing representatives of the *Vesiculovirus* and *Lyssavirus* genera of the *Rhabdoviridae*, but proved to be quite distinct from either one. The available G and M protein sequences of the rhabdoviruses allowed a more detailed phylogenetic analysis, and weighted maximum parsimony analysis of G proteins resulted in a single most parsimonious tree. Similar results were obtained by analysis of the M protein data, and bootstrap analysis indicated that the main branch points on the trees were well supported (Morzunov et al. 1995). The authors concluded that IHNV, VHSV and HIRRV represent a separate cluster within the rhabdoviruses and proposed the creation of a new genus. As a consequence, the new *Novirhabdovirus* genus was established in the seventh report of the ICTV (Walker et al. 2000).

Phenotypic and genetic diversity among IHNV isolates from different geographical sources has been investigated since the 1980s using different methods including protein electropherotyping (Hsu et al. 1986) and monoclonal antibody reactivity (La Patra et al. 1995; Ristow and Arnzen 1989; Winton et al. 1988). Oshima et al. (1995) used a RNase T1 fingerprint analysis to estimate the level of genetic variation among 26 IHNV isolates from salmon and trout from the enzootic area of western North America. All of the isolates contained less than 50% variation in fingerprint spot location, and represented a single fingerprint group. However, sufficient variation was detected to separate the isolates into four subgroups, which appeared to correlate with the different geographic origin.

After the glycoprotein gene sequence of IHNV became available (Koener et al. 1987), genomic sequences were increasingly used for phylogenetic and comparative investigations. Nichol et al. (1995) determined the G and NV sequences of 12 different IHNV isolates to examine the molecular epidemiology and evolution of IHNV. It was shown that both genes were highly conserved with a nucleotide divergence of only 3.6% and 4.4%, respectively. Moreover, the phylogenetic relationship correlated with the geographic origin of virus isolates rather than with host species or time of isolation.

A more comprehensive analysis of genetic diversity and epidemiology of IHNV especially in Alaska was performed by Emmenegger et al. (2000). These authors analyzed 42 IHNV strains isolated from 1976 to 1996 by ribonuclease protection assay (RPA) of the N, G and NV genes and sequencing of part of the G gene and the complete NV gene. The absence of any dominant G gene haplotype in the RPA supported the hypothesis that the G gene maintains a higher level of genetic diversity than other structural IHNV genes. This may be due to an increased immunological selective pressure on this major antigen. A pair-wise comparison of the G and NV sequence data showed that the maximum nucleotide diversity was 2.75% for the NV gene and 1.99% for a 301-bp region of the G gene. A genetic characterization of IHNV isolated from coastal salmonid stocks in Washington State (Emmenegger and Kurath 2002) confirmed the low intrapopulational nucleotide diversity observed in Alaska (Emmenegger et al. 2000). These and other studies document that the genetic diversity of IHNV within Alaska, British Columbia and Washington State is notably lower than in the more southern portions of its North American range (Garver et al. 2003; Troyer et al. 2000, 2003; Nichol et al. 1995).

Recently, the most complete analysis of IHNV genetic diversity and phylogeny in relation to geographical origin was published (Kurath et al. 2003). This study on 323 IHNV field isolates, including single isolates from Japan, France and Italy, was based on sequence analysis of a 303 nucleotide variable region within the glycoprotein gene. It revealed a maximum nucleotide diversity of 8.6%, indicating an overall low genetic diversity. Phylogenetic analysis resulted in the assignment into three major genogroups, designated U, M and L, which varied in geographical range. Levels of genetic diversity within the genogroups indicated that genogroup M, comprising mainly isolates from Idaho trout farms, was three- to fourfold more diverse than the other genogroups (Troyer et al. 2003). Thus, the M genogroup apparently underwent a relatively rapid evolution. In contrast, for the isolates of the U genogroup (mainly of Alaska, British Columbia, Washington coast and Columbia River Basin origin), evolutional stasis was observed (Kurath et al. 2003). Possible hypotheses for this observation are the relatively recent Alaskan aquaculture program and the strict guidelines on the transfer of fish stocks within the state. Furthermore, healthy wild salmon stocks in Alaska do not inflict any exogenous infectious pressure, which combined with the less intense aquacultural activity may also contribute to the low variability of

IHNV isolates (Emmenegger et al. 2000). Different oceanographic features in the enzootic areas together with a possible marine reservoir and a carrier-state for IHNV could also play a role in the transmission of different virus strains to salmonids in saltwater (St Hilaire et al. 2001; Meyers 1998; Kent et al. 1998; Traxler et al. 1997).

The incorporation of three non-American isolates in the phylogenetic studies (Kurath et al. 2003) supported previous hypotheses regarding the spread of IHNV from North America to Europe and Japan (Sano et al. 1977b). An isolate from Japan had a G sequence identical to six other U genogroup isolates from Alaska and British Columbia. The French and Italian isolates fell into the M genogroup but outside the established subgroups, suggesting a common M genogroup ancestor. Based on the overall low genetic diversity and the low mutation frequency of IHNV (Emmenegger et al. 2003), future IHNV isolates will most likely also fall into one of the described three genogroups.

4.2 Hirame Rhabdovirus

Hirame rhabdovirus (HIRRV), a fish rhabdovirus closely related to IHNV, was first isolated in Japan from moribund hirame (*Paralichthys olivaceus;* common names for this fish species are Japanese flounder, bastard halibut, live flounder or false halibut; Kimura et al. 1986). However, other fish species are also infected by HIRRV such as ayu (*Plecoglossus altivelis*), black seabream (*Milio macrocephalus*), and mebaru (*Sebastes inermis*) (Sano and Fukada 1987; Yoshimizu et al. 1987). HIRRV is widely distributed in Japan, and has never been found outside of Japan. First analyses of the genome size and expressed viral proteins were performed by Nishizawa et al. (1991a, 1991b). Serological investigations showed that HIRRV is distinguishable from other fish rhabdoviruses by neutralization tests (Kimura et al. 1986).

The genetic analysis of HIRRV started in 1995 when Nishizawa et al. (1995) sequenced the phosphoprotein and matrix protein genes, and compared the data with the respective IHNV and VHSV sequences. The sequence similarities of the HIRRV and IHNV phosphoproteins were 61.5% at the nucleotide level and 81.5% at the amino acid level, whereas the similarities between HIRRV and three VHSV isolates were 40.5% at the nucleotide level and 44.6%–45.5% at the amino acid level. Similar sequence differences were found for the matrix proteins of HIRRV, IHNV and VHSV. These data indicated that HIRRV is more closely related to IHNV than to VHSV, and that HIRRV is a distinct virus and not an isolate of IHNV. The relationship between HIRRV and IHNV was confirmed by Björklund et al. (1996) who sequenced the glycoprotein gene and the gene junctions of HIRRV. The gene junctions of HIRRV and IHNV are well conserved, and phylogenetic studies based on all available G genes of rhabdoviruses also identified a close relationship between HIRRV and IHNV. Together with SHRV and VHSV, HIRRV and IHNV cluster in a separate genus within the *Rhabdoviridae* family. As a prerequisite for a membership within the *Novirhabdovirus* genus, HIRRV specifies a NV protein of 111 amino acids. It is identical in size to the NV protein of IHNV, whereas the NV protein of VHSV contains 122 amino acids. The NV protein sequence identity between HIRRV and IHNV amounts to 54.1%, whereas the identity to VHSV reaches only 16.5% (Kurath et al. 1997). In 1998, a complete sequence of the HIRRV isolate CA9703 was submitted to GenBank (Oh and Choi, unpublished, accession number AF 104985), which was updated in September 2003 (Oh et al., unpublished, accession number NC_005093). A size of the HIRRV genome of 11,034 bp was established and the gene order 3′-N-P-M-G-NV-L-5′ was confirmed. Take together published phylogenetic studies and sequence comparisons of HIRRV as well as the phylogenetic tree depicted in this report (Fig. 1) clearly support the assignment of HIRRV to the *Novirhabdovirus* genus.

4.3
Viral Hemorrhagic Septicemia Virus

Viral hemorrhagic septicemia (VHS), or Egtved disease, is one of the most serious viral diseases in farmed salmonid fishes (Wolf 1988). The clinical symptoms comprise acute, chronic and nervous forms of the disease. Acute signs are typically associated with a rapid onset of heavy mortality. Fish are lethargic and anemic, and exhibit dark exophthalmus. Hemorrhages are evident in the eyes, skin, and gills and at the bases of the fins.

The causative agent, VHSV, infects a variety of freshwater and marine fish (Dixon et al. 2003; Hedrick et al. 2003; Dopazo et al. 2002; Brudeseth et al. 2002; King et al. 2001; Smail 2000; Hershberger et al. 1999; Meyers et al. 1999; Mortensen et al. 1999; Castric et al. 1992), causing a high mortality among farmed rainbow trout (*Oncorhynchus mykiss)* throughout Europe (de Kinkelin et al. 1979). An extensive pool of enzootic viruses exists in fish in continental Europe, where the disease was first described in 1938 (Schäperclaus 1938, 1990). Subsequently, regular appearances have been observed since the 1950s. The first isolation of marine VHSV was from Atlantic cod (*Gadus morhua*) (Jensen et al. 1979), and subsequent outbreaks were reported in sea-farmed rainbow trout in France (Castric and de Kinkelin 1980), and turbot (*Scophthalmus maximus*) in Germany and Scotland (Ross et al. 1994; Schlotfeldt et al. 1991). In the late 1980s, VHSV was for the first time isolated in western North America from healthy adult Coho salmon (*Oncorhynchus kisutch*) and from diseased Pacific cod (*Gadus macrocephalus*) (reviewed in Meyers and Winton 1995; Brunson et al. 1989; Winton et al. 1989). More recently in 1999, VHSV was first isolated in Japan from wild Japanese flounder (*Paralichthys olivaceus*) during a survey on the distribution of viruses in wild marine fish (Takano et al. 2000, 2001). Immediately thereafter, VHSV infections occurred in farmed Japanese flounder in the Seto Inland Sea of Japan (Isshiki et al. 2001).

Cross-neutralization analyses using strain-specific polyclonal and monoclonal antibodies identified between three and four reactivity patterns (Olesen et al. 1993; Castric et al. 1992; Le Berre et al. 1977; Vestergard-Jorgensen 1972) indicative of intrapopulational differences of VHSV. Oshima et al. (1993) compared eight isolates from North America and Europe by T1 ribonuclease fingerprinting analysis and reported that the isolates from the two continents segregated into two different fingerprint groups. Moreover, they observed patterns of oligonucleotide spots, which showed some differences on the basis of the host fish species. Later investigations based on VHSV sequences confirmed a pattern of sequence variations related to the host species (Benmansour et al. 1997; Basurco et al. 1995).

At the end of the 1980s the molecular analysis of VHSV was increasingly based on sequence data of the viral genome. Analyses of the N, P, M and G genes showed that VHSV is closely related to IHNV (Benmansour et al. 1994; Bernard et al. 1990, 1992; Thiry et al. 1991). The analysis of the gene junction between the glycoprotein and the polymerase protein genes identified a sixth open-reading frame encoding a non-virion protein NV (Schütze et al. 1996; Basurco and Benmansour 1995). The first complete genomic sequence of VHSV confirmed the presence of the NV protein gene and verified that VHSV is related to IHNV (Schütze et al. 1999). Thus, classification of VHSV into the genus *Novirhabdovirus* was formally accepted (Walker et al. 2000). In addition, comparison of the VHSV genome and gene products with other rhabdoviruses revealed a high homology to those of IHNV with amino acid identities of between 37% and 60%. Only the NV protein exhibits a lower identity of only 23%.

Phylogenetic analyses based on the nucleotide sequence data demonstrated that VHSV isolates may be grouped according to their geographic origin rather than to their host species. Studies based on the glycoprotein gene of VHSV identified geographically distinct genogroups comprising isolates from North America (genogroup 3), the European marine environment (genogroup 2) and continental Europe (genogroup 1) (Benmansour et al. 1997; Stone et al. 1997). The genetic divergence between the North American and the European strains in the glycoprotein gene is 15%–18% and thus, a differentiation by RT-PCR is possible (Einer-Jensen et al. 1995). Further analysis of European marine isolates with RNase protection assay and sequence data of the nucleoprotein gene allowed subdivision of the European marine isolates into two subgroups circulating in the North and the Baltic Sea, respectively. In addition, marine isolates from the Baltic Sea were identified that are closely related to isolates from freshwater aquaculture in continental Europe (Snow et al. 1999).

VHSV strains isolated from coastal areas of western Japan and from the Seto Inland Sea of Japan (Takano et al. 2000, 2001; Isshiki et al. 2001) could also be classified into the major genogroups (Nishizawa et al. 2002). Most Japanese isolates clustered within the North American group (genogroup 3), and only one isolate was similar to strains of genogroup 1 (traditional continental Europe isolates). Since the VHSV isolates from Japanese coastal areas formed a separate cluster within genogroup 3, the authors suggested that VHS may be enzootic in Japanese waters (Nishizawa et al. 2002). The single isolate of genogroup 1 was probably introduced to Japan from Europe.

A phylogenetic study using VHSV isolated in France from 1971 to 1999 confirmed the good correlation between the geographical origin of the isolates and their genetic characteristics (Thiery et al. 2002). Most of the VHSV isolates clustered in genogroup 1, which contains the VHSV strains isolated in continental Europe. Only one isolate fell into genogroup 2, the cluster containing marine European VHSV isolates. Interestingly, this isolate was derived from wild eel *Anguilla anguilla* caught in the river Loire estuary (Castric et al. 1992), showing that this migrating fish can also be naturally infected by genogroup 2 VHSV.

In addition, sequence analyses of nearly the complete genomes of two virulent freshwater and two avirulent marine VHSV strains revealed greater than 97.2% nucleotide sequence identity and greater than 98.6% amino acid similarity, confirming the close relationship between marine and freshwater strains (Betts and Stone 2000) and demonstrating that only a limited number of amino acid residues may be involved in the determination of VHSV virulence for salmonids. Thus, marine VHSV strains could represent a potential risk for aquaculture. Otherwise, it was reported that after in vivo passages of marine VHSV under stable conditions, no sequence variations in the glycoprotein gene were observed (Snow and Cunningham 2000).

In summary, the increased number of VHSV isolations from marine fish, together with the established susceptibility of numerous marine and freshwater species to this virus, led to a reconsideration of the epidemiology of VHSV. First, it was suggested that the reservoir of VHSV was located in continental Europe, because VHS was described since the 1930s in Europe, and VHSV had been regularly isolated there since the 1960s. However, VHSV has been isolated from marine fish since the late 1970s, and these viruses were mostly avirulent for salmonid fish (King et al. 2001; Dixon et al. 1997; Kocan et al, 1997; Castric and de Kinkelin 1980). Stone et al. (1997) suggested that all marine fish may be potential VHSV carriers since the virus has often been isolated from Atlantic and Pacific fish. The occurrence of very low virus titers in these fish support the hypothesis that they may act as carriers (Dopazo et al. 2002). Along these lines, it has been suggested that VHSV had been enzootic in the Pacific Ocean and European waters for quite some time, and that the viruses observed in mainland Europe may have a marine origin (Stone et al. 1997). The highly pathogenic nature of the European freshwater strains of VHSV may then be a result of both the exposure of rainbow trout to the stressful environment of intensive culture predisposing them for disease and the high rate of mutation inherent in all rhabdoviruses resulting in more virulent viruses (Meyers and Winton 1995).

4.4 Snakehead Rhabdovirus

Snakehead rhabdovirus (SHRV) was isolated from a diseased snakehead (*Ophicephalus striatus*) in Thailand. The disease was characterized by necrotic ulcerations and affected both wild and cultured fish throughout South-east Asia. Although several fish species were infected, it was most prevalent in cultured snakehead (Kasornchandra et al. 1991). SHRV was one of several pathogens associated with epizootic ulcerative syndrome (EUS) already described above (Kasornchandra et al. 1992; Frerichs et al. 1986; Wattanavijarn et al. 1986). Initial characterization of SHRV using electron microscopy, protein profiles and cross-neutralization tests indicated a close relationship to IHNV, HIRRV and VHSV, supporting an assignment to the *Novirhabdovirus* genus (Kasornchandra et al. 1992). Surprisingly, SHRV is able to replicate at temperatures as high as 35°C (optimal between 28°C and 31°C), whereas all of the previously identified novirhabdoviruses replicate well only below 20°C.

The first sequence analyses and phylogenetic studies of SHRV were published by Johnson et al. (1999). The authors sequenced the glycoprotein gene and compared the nucleotide and deduced protein sequences with other rhabdoviruses. SHRV-G was found to possess 47%, 37%, and 36% amino acid identity to VHSV-G, IHNV-G, and HIRRV-G, respectively. The phylogenetic analysis confirmed the close relationship of SHRV to the novirhabdoviruses, especially to VHSV (Johnson et al. 1999).

In Septemper 1999, the complete sequence of the 11,550-bp genome of SHRV was submitted to GenBank (Accession number AF147498) and was updated in August 2003 (Accession number NC_000903). Based on this sequence data, Johnson et al. (2000) generated a recombinant SHRV with an unfunctional NV protein. Recombinant SHRV was obtained after T7 RNA polymerase transcription of a full-length genomic clone in EPC cells using a recombinant vaccinia virus system. Although SHRV is not the prototype novirhabdovirus, it is better suited for reverse genetics experiments than, for example, IHNV or VHSV, as it is the only member of the *Novirhabdovirus* genus that can replicate at temperatures that favor vaccinia virus infection and efficient T7 RNA polymerase transcription. Recombinant SHRV appeared to be morphologically identical to the wild-type, and the engineered NV knockout virus was produced at a titer as high as that of wild-type virus in cultured fish cells. Thus, the NV protein was demonstrated to be non-essential for replication of SHRV. Since these findings are in contrast to results obtained for IHNV (Biacchesi et al. 2000), the importance of the NV protein remains unclear. Identity of the NV gene sequence of SHRV with NV sequence data of IHNV and VHSV are 37% and 40%, respectively. Again, the NV genes are more divergent compared to the structural genes. Identities of the N, P, M, G and L genes of SHRV and IHNV amount to 50%, 45%, 47%, 49%, and 58%, respectively, while comparison of the N, P, M, G and L genes between SHRV and VHSV shows 56%, 52%, 51%, 55% and 63% identity. Thus, the identities of all genes are higher for SHRV and VHSV than for IHNV.

In the seventh report of the ICTV, SHRV has been classified as a tentative species in the *Novirhabdovirus* genus (Walker et al. 2000). The results of sequence analyses with the identification of an NV gene, as well as the phylogenetic studies of Johnson et al. (1999) and the data presented in this report (Fig. 1) clearly support the assignment of SHRV to the *Novirhabdovirus* genus.

4.5
General Remarks for the *Novirhabdovirus* Genus

Classification of the *Novirhabdovirus* genus is in primarily based on the presence of the non-virion gene located between the glycoprotein and the polymerase genes. However, comparison of nucleotide and amino acid sequences of the structural proteins also support the classification of the novirhabdoviruses into a separate genus (Fig. 1). The amino acid sequences of the NV protein are significantly less well conserved. For instance, there is no significant amino acid sequence homology between the NV protein of IHNV and VHSV (Kurath et al. 1997). It is possible that the role of the NV protein, which is still unknown, is different in the different novirhabdoviruses. Whatever this possible divergent role is, it is unlikely to involve a host specific phenomenon, since the natural hosts for both IHNV and VHSV are salmon and trout. The successful recovery of recombinant IHN virus expressing foreign genes instead of the NV gene demonstrated that the NV protein was not required for viral replication in cell culture, although its presence greatly improved virus growth (Biacchesi et al. 2000). In contrast, a NV knockout mutant of SHRV showed no impairment of virus propagation and replicated in vitro to the same virus titer as the wild-type *SHRV* (Johnson et al. 2000).

Only a few aquatic viruses have been analyzed in comprehensive phylogeographical studies up to now. The best example to date are the investigations of VHSV, which has diverged into three to four genogroups that correlate with geographical origin (Snow et al. 1999; Stone et al. 1997; Benmansour et al. 1997). Similar results were found for IHNV. However, different intrapopulational sequence diversities were observed for these two virus species. A maximum of 15%–18% nucleotide diversity was found between European and North American VHSV isolates using full-length G sequences. In comparison, the maximum nucleotide diversity reported for G gene sequences of rabies virus (genotype 1) and vesicular stomatitis virus is 16%–20% (Tordo et al. 1993; Bilsel and Nichol 1990; Nichol et al. 1989). Thus, with a maximum of only 8.6% nucleotide diversity in the variable mid-G region, the overall genetic diversity of IHNV is low compared with other rhabdoviruses. To date all IHNV isolates fall into three major genogroups, designated U, M and L. Interestingly, the M genogroup exhibits a significantly higher diversity than the other two genogroups (Kurath et al. 2003; Troyer et al. 2003; Garver et al. 2003; Emmenegger and Kurath 2002; Emmenegger et al. 2000).

Sequence data as well as phylogenetic analyses of HIRRV and SHRV clearly identified both viruses as members of the *Novirhabdovirus* genus. HIRRV is more related to IHNV, whereas SHRV exhibits a closer genetic relationship to VHSV. Based on the requirements for assignment of viruses into the *Novirhabdovirus* genus, we propose the classification of SHRV as an official member of the genus.

Although so far only aquatic novirhabdoviruses have been detected, it is interesting to speculate whether mammalian rhabdoviruses could be found that also specify an NV gene. When the function and importance of the NV protein for the novirhabdoviruses is known, a selective and more intensive search for novel novirhabdoviruses based on these results may be possible.

5
Concluding Remarks

Modern molecular epidemiology can distinguish and classify virus strains based on genetic differences. This information can identify links between infections, aid in determining the modes of virus strain transmission, and provide estimates of the rate of virus evolution through space, time, and within a host species. A pioneering technique to genetically characterize fish virus genomes was the ribonuclease protection assay, which produces a genetic fingerprint (RPA) of the genome (Emmenegger et al. 2000). The advantage of RPA is its ability to rapidly analyze large regions of the genome of many isolates. A limitation of the RPA is that not all nucleotide differences between isolates are detected (Meyers et al. 1985; Winter et al. 1985). Especially for investigations of mutations within a geographic area or individual hosts, genome sequencing is the method of choice. Genetic characterization of multiple isolates from separate sites can clarify whether viruses are evolving and/or whether exogenous sources account for the presence of different strains in an area. Both viral evolution and viral introduction can contribute significantly to the heterogeneity of viruses (Emmenegger et al. 2000).

Rhabdoviruses are important pathogens in farmed and wild fish. The family contains numerous previously unclassified and unassigned rhabdoviruses, such as *viral hemorrhagic septicemia virus* (VHSV) (Jensen 1963), *infectious hematopoietic necrosis virus* (IHNV) (Amend et al. 1969), *spring viremia of carp virus* (SVCV) (Fijan et al 1971), *pike fry rhabdovirus* (PFRV) (de Kinkelin et al. 1973), *eel virus American* (EVA) (Sano 1976), *eel virus European* X (EVEX) (Sano et al. 1977a), cod ulcus-syndrome rhabdovirus (Jensen et al. 1979), Rio Grande perch rhabdovirus (Malsberger and Lautenslager 1980), rhabdovirus salmonis (Osadchaya and Nakonechnaya 1981), perch rhabdovirus (Dorson et al. 1984), *hirame rhabdovirus* (HIRRV) (Kimura et al. 1986), *ulcerative disease rhabdovirus* (UDRV) (Frerichs et al. 1986), *snakehead rhabdovirus* (SHRV) (Kasornchandra et al. 1992), *Trout rhabdovirus* 903/87 (TRV-903/87) (Koski et al. 1992), pike rhabdovirus DK 5533 (Jorgensen et al. 1993), *sea trout rhabdovirus* 28/97 (STRV-28/97) (Björklund et al. 1994) and Chinese sucker rhabdovirus (Zhang et al. 2000) isolated from cultured fish. Rhabdoviruses such as SVCV (Johnson et al. 1999) was also isolated from non-vertebrates. Most fish rhabdoviruses have been analyzed in terms of their morphology, cytopathogenicity, serological relatedness, structural polypeptide composition, and electrophoretic migration pattern of the viral structural proteins in an SDS-PAGE (Granzow et al. 1997; Björklund et al. 1994; Lilley and Frerichs 1994; Kasornchandra et al. 1992; Frerichs 1986; Ahne et al. 1988). The development of new methods for fast and efficient virus genome characterization such as reverse transcriptase-polymerase chain reaction (RT-PCR) and nucleotide sequencing as well as the processing of the produced data by new bioinformatic programs has drastically increased the speed and reliability of genome-wide screens. These new techniques permit an in-depth comparison of nucleotide and protein sequences of the fish rhabdoviruses between themselves and with the other members of the *Rhabdoviridae* family. To date all fish rhabdoviruses are classified in two genera within the family *Rhabdoviridae*. Members of the *Novirhabdovirus* genus contain an additional NV (non-virion) gene between the glycoprotein and the polymerase genes. Phylogenetic analyses based on sequence information substantiated the creation of this new genus (Walker et al. 2000). The situation within the *Vesiculovirus* genus is less clear. Although the fish rhabdoviruses SVCV, PFRV, TRV-903/87 and STRV-28/97 clustered within the *Vesiculovirus* genus, the rather low homologies of these fish viruses with the mammalian vesiculoviruses are remarkable. Further sequence analyses of additional virus isolates from different areas throughout the world could help to establish a definite assignment of the vesiculo-type fish rhabdoviruses within the *Rhabdoviridae* family. In the future, the rapid development of molecular technologies such as nucleotide sequencing and gene chip technique could be utilized to perform high-speed whole genome analyses. It has been further pointed out that due to the absence of intergenomic recombination in rhabdoviruses, results of molecular epidemiological studies should be essentially independent of the genomic site analyzed (Nadin-Davies 2000), provided the analyses of distinct sequence data are sufficiently robust. However, the identification of highly variable genome regions for a more detailed determination of the molecular epidemiology of fish rhabdoviruses as well as the delineation of virological markers and factors are important tasks for determination of the association between causative agent and disease.

References

Ahne W (1975) A rhabdovirus isolated from grass carp (*Ctenopharyngodon idella* Val.). Arch Virol 48:181–185

Ahne W, Bjorklund HV, Essbauer S, Fijan N, Kurath G, Winton JR (2002) Spring viremia of carp (SVC). Dis Aquat Organ 52: 261–272

Ahne W, Jorgensen PE, Olesen NJ, Wattanavijarn W (1988) Serological examination of a rhabdovirus isolated from snakehead (*Ophicephalus striatus*) in Thailand with ulcerative syndrome. J Appl Ichthyol 4:194–196

Ahne W, Mahnel H, Steinhagen P (1982) Isolation of pike fry rhabdovirus from tench, *Tinca tinca* L., and white bream, *Blicca bjoerkna* (L). J Fish Dis 5:535–537

Ahne W, Thomsen I (1986) Isolation of pike fry rhabdovirus from *Pseudorasbora parva*. J Fish Dis 9:555–556

Amend DF, Yasutake WT, Mead RW (1969) A haematopoietic virus disease of rainbow trout and sockeye salmon. Trans Am Fish Soc 98:796–804

Basurco B, Benmansour A (1995) Distant strains of the fish rhabdovirus VHSV maintain a sixth functional cistron which codes for a nonstructural protein of unknown function. Virology 212:741–745

Basurco B, Vende P, Monnier AF, Winton JR, de Kinkelin P, Benmansour A (1995) Genetic diversity and phylogenetic classification of viral hemorrhagic septicemia virus (VHSV). Vet Res 26:460–463

Benmansour A, Basurco B, Monnier AF, Vende P, Winton JR, de Kinkelin P (1997) Sequence variation of the glycoprotein gene identifies three distinct lineages within field isolates of viral haemorrhagic septicaemia virus, a fish rhabdovirus. J Gen Virol 78:2837–2846

Benmansour A, Paubert G, Bernard J, De Kinkelin P (1994) The polymerase-associated protein (M1) and the matrix protein (M2) from a virulent and an avirulent strain of viral hemorrhagic septicemia virus (VHSV), a fish rhabdovirus. Virology 198:602–612

Bernard J, Bremont M (1995) Molecular biology of fish viruses: a review. Vet Res 26:341–351

Bernard J, Bremont M, Winton J (1992) Nucleocapsid gene sequence of a North American isolate of viral haemorrhagic septicaemia virus, a fish rhabdovirus. J Gen Virol 73:1011–1014

Bernard J, Lecocq-Xhonneux F, Rossius M, Thiry ME, de Kinkelin P (1990) Cloning and sequencing the messenger RNA of the N gene of viral haemorrhagic septicaemia virus. J Gen Virol 71:1669–1674

Betts AM, Stone DM (2000) Nucleotide sequence analysis of the entire coding regions of virulent and avirulent strains of viral haemorrhagic septicaemia virus. Virus Genes 20:259–262

Biacchesi S, Thoulouze MI, Bearzotti M, Yu YX, Bremont M (2000) Recovery of NV knockout infectious hematopoietic necrosis virus expressing foreign genes. J Virol 74:11247–1153

Bilsel PA, Nichol ST (1990) Polymerase errors accumulating during natural evolution of the glycoprotein gene of vesicular stomatitis virus Indiana serotype isolates. J Virol. 64:4873–4883

Björklund HV, Emmenegger EJ, Kurath G (1995) Comparison of the polymerases (L genes) of spring viremia of carp virus and infectious hematopoietic necrosis virus. Vet Res 26:394–398

Björklund HV, Higman KH, Kurath G (1996) The glycoprotein genes and gene junctions of the fish rhabdoviruses spring viremia of carp virus and hirame rhabdovirus: analysis of relationships with other rhabdoviruses. Virus Res 42:65–80

Björklund HV, Olesen NJ, Jorgensen PEV (1994) Biophysical and serological characterization of rhabdovirus 903/87 isolated from European lake trout *Salmo trutta lacustris*. Dis Aquat Organ 19:21–26

Bootland LM, Leong JC (1999) Infectious hematopoietic necrosis virus. In: Woo PTK, Bruno DW (eds) Fish diseases and disorders. Vol. 3. CAB International, Wallingford, UK, pp 57–112

Bourhy H, Kissi B, Tordo N (1993) Molecular diversity of the *Lyssavirus* genus. Virology 194:70–81

Bovo G, Giorgetti G, Jorgensen PEV, Olesen NJ (1987) Infectious haematopoietic necrosis: first detection in Italy. Bull Eur Ass Fish Path 7:124

Brudeseth BE, Evensen O (2002) Occurrence of viral haemorrhagic septicaemia virus (VHSV) in wild marine fish species in the coastal regions of Norway. Dis Aquat Organ 52:21–28

Brunson R, True K, Yancey J (1989) VHS virus isolated at Makah national fish hatchery. Am Fish Soc Fish Health Sect Newl 17:3–4

Castric J, de Kinkelin P (1980) Occurrence of viral haemorrhagic septicaemia in rainbow trout *Salmo gairdneri* Richardson reared in seawater. J Fish Dis 3:21–27

Castric J, Jeffroy J, Bearzotti M, de Kinkelin P (1992) Isolation of viral haemorrhagic septicaemia virus (VHSV) from wild elvers *Anguilla anguilla*. Bull Eur Assoc Fish Pathol 12:21–23

Choi TJ, Kuwata S, Koonin EV, Heaton LA, Jackson AO (1992) Structure of the L (polymerase) protein gene of sonchus yellow net virus. Virology 189:31–39

De Kinkelin P, Chilmonczyck S, Dorson M, Le Berre M, Baudouy AM (1979) Some pathogenic facets of rhabdoviral infection of salmonid fish. In: Bachmann PA (ed) Munich Symposia on Microbiology: mechanisms of viral pathogenesis and virulence. Munich, WHO Collaborating Centre, pp 357–375

De Kinkelin P, Galimard B, Bootsma R (1973) Isolation and identification of the causative agent of 'red disease' of pike (*Esux lucius* L. 1766). Nature 241:465–467

Dixon PF, Avery S, Chambers E, Feist S, Mandhar H, Parry L, Stone DM, Strommen HK, Thurlow JK, Lui CT, Way K (2003) Four years of monitoring for viral haemorrhagic septicaemia virus in marine waters around the United Kingdom. Dis Aquat Organ 54:175–186

Dixon PF, Feist S, Kehoe E, Parry L, Stone DM,Way K (1997) Isolation of viral haemorrhagic septicaemia virus from Atlantic herring *Clupea harengus* from the English Channel. Dis Aquat Organ 30:81–89

Dixon PF, Hattenberger-Baudouy AM, Way K (1994) Detection of carp antibodies to spring viremia of carp virus by competitive immunoassay. Dis Aquat Org 19:181–186

Dopazo CP, Bandin I, Lopez-Vazquez C, Lamas J, Noya M, Barja JL (2002) Isolation of viral hemorrhagic septicemia virus from Greenland halibut *Reinhardtius hippoglossoides* caught at the Flemish Cap. Dis Aquat Organ 50:171–179

Dorson M, Torchy C, Chilmonczyk S, de Kinkelin P, Michel C (1984) A rhabdovirus pathogenic for perch, *Perca fluviatilis* L.: isolation and preliminary study. J Fish Dis 7:241–245

Einer-Jensen K, Olesen NJ, Lorenzen N, Jorgensen PE (1995) Use of the polymerase chain reaction (PCR) to differentiate serologically similar viral haemorrhagic septicaemia (VHS) virus isolates from Europe and America. Vet Res 26:464–469

Emmenegger EJ, Kurath G (2002) Genetic characterization of infectious hematopoietic necrosis virus of costal salmonid stocks in Washington state. J Aquat Anim Health 14:25–34

Emmenegger EJ, Meyers TR, Burton TO, Kurath G (2000) Genetic diversity and epidemiology of infectious hematopoietic necrosis virus in Alaska. Dis Aquat Organ 40:163–76

Emmenegger EJ, Troyer RM, Kurath G (2003) Characterization of the mutant spectra of a fish RNA virus within individual hosts during natural infections. Virus Res 96:15–25

Fijan N, Petrinec Z, Sulimanovic D, Zwillenberg LO (1971) Isolation of the causative agent from the acute form of infectious dropsy of carp. Veterinarski Archiv (Zagreb) 41:125–138

Fijan N, Matasin Z, Jeney Z, Oláh J, Zwillenberg LO (1984) Isolation of Rhabdovirus carpio from sheatfish (*Silurus glanis*) fry. In: Olày J (ed) Fish, pathogens and environment in European polyculture. Akadémiai Kiadó, Budapest, pp 17–24

Frerichs GN (1995) Viruses associated with the epizootic ulcerative syndrome (EUS) of fish in South-East Asia. Vet Res 26:449–454

Frerichs GN, Millar SD, Roberts RJ (1986) Ulcerative rhabdovirus in fish in South-East Asia. Nature 322:216

Garver KA, Troyer RM, Kurath G (2003) Two distinct phylogenetic clades of infectious hematopoietic necrosis virus overlap within the Columbia River basin. Dis Aquat Organ 55:187–203

Granzow H, Weiland F, Fichtner D, Enzmann PJ (1997) Studies of the ultrastructure and morphogenesis of fish pathogenic viruses grown in cell culture. J Fish Dis 20:1–10

Grischkowsky RS, Amend DF (1976) Infectious hematopoietic necrosis virus, prevalence in certain Alaska sockeye salmon, *Oncorhynchus nerka*. J Fish Res Board Can 33:186–188

Guenther RW, Watson SW, Rucker RR (1959) Etiology of sockeye salmon "virus" disease. U S Fish Wildl Serv Spec Sci Rep Fish 296:1–10

Haenen OL, Davidse A (1989) Isolation of pike fry rhabdovirus from roach (*Rutilus rutilus*). Bull Eur Assoc Fish Pathol 9:116

Heaton LA, Hillman BI, Hunter BG, Zuidema D, Jackson AO (1989) Physical map of the genome of sonchus yellow net virus, a plant rhabdovirus with six genes and conserved gene junction sequences. Proc Natl Acad Sci U S A 86: 8665–8668

Hedrick RP, Batts WN, Yun S, Traxler GS, Kaufman J, Winton JR (2003) Host and geographic range extensions of the North American strain of viral hemorrhagic septicemia virus. Dis Aquat Organ 55:211–220

Hershberger PK, Kocan RM, Elder NE, Meyers TR, Winton JR (1999) Epizootiology of viral hemorrhagic septicemia virus in Pacific herring from the spawn-on-kelp fishery in Prince William Sound, Alaska, USA. Dis Aquat Organ 37:23–31

Hill BJ (1975) Physico-chemical and serological characterization of five rhabdoviruses infecting fish. J Gen Virol 27:369–378

Hillis DM, Bull JJ (1993) An empirical test of bootstrapping as a method for assessing confidence in phylogenetic analysis. Syst Biol 42:182–192

Hoffmann B, Schütze H, Mettenleiter TC (2002) Determination of the complete genomic sequence and analysis of the gene products of the virus of spring viremia of carp, a fish rhabdovirus. Virus Res 84:89–100

Hsu YL, Engelking HM, Leong JC (1986) Occurrence of different types of infectious hematopoietic necrosis virus in fish. Appl Environ Microbiol 52:1353–1361

Isshiki T, Nishizawa T, Kobayashi T, Nagano T, Miyazaki T (2001) An outbreak of VHSV (viral hemorrhagic septicemia virus) infection in farmed Japanese flounder *Paralichthys olivaceus* in Japan. Dis Aquat Organ 47:87–99

Jensen MH (1963) Preparation of fish tissues for virus research. Bull Off Int Epizoot 59:131–134

Jensen NJ, Bloch B, Larsen JL (1979) The ulcus-syndrome in cod (*Gadus morhua*): III. A preliminary virological report. Nord Vetmed 31:436–442

Johansson T, Nylund S, Olesen NJ, Bjorklund H (2001) Molecular characterisation of the nucleocapsid protein gene, glycoprotein gene and gene junctions of rhabdovirus 903/87, a novel fish pathogenic rhabdovirus. Virus Res 80:11–22

Johansson T, Ostman-Myllyoja L, Hellstrom A, Martelius S, Olesen NJ, Bjorklund H (2002) A novel fish rhabdovirus from Sweden is closely related to the Finnish rhabdovirus 903/87. Virus Genes 25:127–138

John KR, George MR, Richards RH, Frerichs GN (2001) Characteristics of a new reovirus isolated from epizootic ulcerative syndrome infected snakehead fish. Dis Aquat Organ 46:83–92

Johnson MC, Maxwell JM, Loh PC, Leong JA (1999) Molecular characterization of the glycoproteins from two warm water rhabdoviruses: snakehead rhabdovirus (SHRV) and rhabdovirus of penaeid shrimp (RPS)/spring viremia of carp virus (SVCV). Virus Res 64:95–106

Johnson MC, Simon BE, Kim CH, Leong JA (2000) Production of recombinant snakehead rhabdovirus: the NV protein is not required for viral replication. J Virol 74:2343–50

Jorgensen PEV, Olesen NJ, Ahne W, Wahli T, Meier W (1993) Isolation of a previously undescribed rhabdovirus from pike *Esox lucius*. Dis Aquat Organ 16:171–179

Kasornchandra J, Engelking HM, Lannan CJ, Rohovec JS, Fryer JL (1992) Characteristics of three rhabdoviruses from snakehead fish (*Ophicephalus striatus*). Dis Aquat Organ 13:89–94

Kasornchandra J, Lannan CN, Rohovec JS, Fryer JL (1991) Characterization of a rhabdovirus isolated from the snakehead fish (*Ophicephalus striatus*). In: Second International Symposium on Viruses of Lower Vertebrates, pp 175–182

Kent ML, Traxler GS, Kieser D, Richard J, Dawe SC, Shaw RW, Properi-Porta G, Ketcheson J, Evelin TPT (1998) Survey of salmonid pathogens in ocean-caught fishes in British Columbia, Canada. J Aquat Anim Health 10:211–219

Kimura T, Yoshimizu M, Gorie S (1986) A new rhabdovirus isolated in Japan from cultured hirame (Japanese flounder, *Paralichthys olivaceus*) and ayu (*Plecoglossus altivelis*). Dis Aquat Organ 1:209–217

King JA, Snow M, Smail DA, Raynard RS (2001) Distribution of viral haemorrhagic septicaemia virus in wild fish species of the North Sea, northeast Atlantic Ocean and Irish Sea. Dis Aquat Organ 47:81–86

King JA, Snow M, Skall HF, Raynard RS (2001) Experimental susceptibility of Atlantic salmon *Salmo salar* and turbot *Scophthalmus maximus* to European freshwater and marine isolates of viral haemorrhagic septicaemia virus. Dis Aquat Organ 47:25–31

Kiuchi A, Roy P (1984) Comparison of the primary sequence of spring viremia of carp virus M protein with that of vesicular stomatitis virus. Virology 134:238–243

Kocan R, Bradley M, Elder N, Meyers T, Batts W, Winton J (1997) North American strain of viral hemorrhagic septicemia virus is highly pathogenic for laboratory-reared Pacific herring. J Aquat Anim Health 9:278–290

Koener JF, Passavant CW, Kurath G, Leong J (1987) Nucleotide sequence of a cDNA clone carrying the glycoprotein gene of infectious hematopoietic necrosis virus, a fish rhabdovirus. J Virol 61:1342–1349

Koski P, Hill BJ, Way K, Neuvonen E, Rintamäki P (1992) A rhabdovirus isolated from brown trout (*Salmo trutta* m. *lacustris* (L)) with lesions in parenchymatous organs. Bull Eur Ass Fish Pathol 12:177–180

Kurath G, Ahern KG, Pearson GD, Leong JC (1985) Molecular cloning of the six mRNA species of infectious hematopoietic necrosis virus, a fish rhabdovirus, and gene order determination by R-loop mapping. J Virol 53:469–476

Kurath G, Garver KA, Troyer RM, Emmenegger EJ, Einer-Jensen K, Anderson ED (2003) Phylogeography of infectious haematopoietic necrosis virus in North America. J Gen Virol 84: 803–814

Kurath G, Higman KH, Bjorklund HV (1995) The NV genes of fish rhabdoviruses: development of RNase protection assays for rapid assessment of genetic variation. Vet Res 26:477–485

Kurath G, Higman KH, Bjorklund HV (1997) Distribution and variation of NV genes in fish rhabdoviruses. J Gen Virol 78:113–117

Kurath G, Leong JC (1985) Characterization of infectious hematopoietic necrosis virus mRNA species reveals a nonvirion rhabdovirus protein. J Virol 53:462–468

Kurath G, Leong JC (1987) Transcription in vitro of infectious haematopoietic necrosis virus, a fish rhabdovirus. J Gen Virol 68:1767–1771

LaPatra SE, Lauda KA, Jones GR, Walker SC, Shewmaker BS, Morton AW (1995) Characterization of IHNV isolates associated with neurotropism. Vet Res 26:433–437

Laurencin B (1987) IHN in France. Bull Eur Ass Fish Path 7:104

Le Berre M, de Kinkelin P, Metzger A (1977) Identification sérologique des rhabdovirus des salmonidés. Bull Off Intern Epizoo 87:391–393

Leong JC, Hsu YL, Engelking HM, Mulcahy D (1981) Strains of infectious hematopoietic necrosis (IHN) virus may be identified by structural protein differences. In: Karger S (ed) International Symposium Fish Biology: serodiagnostics and vaccines, Leetown, W VA, USA. Develop Biol Standard 49:43–55

Lilley JH, Frerichs GN (1994) Comparison of rhabdoviruses associated with epizootic ulcerative syndrome (EUS) with respect to their structural proteins, cytopathology and serology. J Fish Dis 17:513–522

Lilley JH, Hart D, Panyawachira V, Kanchanakhan S, Chinabut S, Soderhall K, Cerenius L (2003) Molecular characterization of the fish-pathogenic fungus *Aphanomyces invadans*. J Fish Dis 26:263–275

Lio-Po GD, Traxler GS, Albright LJ, Leano EM (2000) Characterization of a virus obtained from snakeheads *Ophicephalus striatus* with epizootic ulcerative syndrome (EUS) in the Philippines. Dis Aquat Organ 43:191–198

Lu YA, Nadala EC Jr, Brock JA, Loh PC (1991) A new virus isolate from infectious hypodermal and hematopoietic necrosis virus (IHHNV)-infected penaeid shrimps. J Virol Methods 31:189–195

Malsberger RG, Lautenslager G (1980) Fish viruses: rhabdovirus isolated from a species of the family *Cichlidae*. Fish Health News 9:1–2

Mastan SA, Qureshi TA (2001) Role of bacteria in the epizootic ulcerative syndrome (EUS) of fishes. J Environ Biol 22(3): 187–92

McAllister PE (1979) Fish viruses and viral infections. In: Fraenkel-Conrat H, Wagner RR (eds) Comprehensive virology 14. , Plenum Press, New York, pp 401–470

Meyers TR (1998) Healthy juvenile sockeye salmon reared in virus-free hatchery water return as adults infected with infectious hematopoietic necrosis virus (IHNV): a case report and review of controversial issues in the epizootiology of IHNV. J Aquat Anim Health 10:172–181

Meyers RM, Larin X, Maniatis T (1985) Detection of single base substitutions by ribonuclease cleavage at mismatches in RNA/DNA duplexes. Science 230:1242–1246

Meyers TR, Short S, Lipson K (1999) Isolation of the North American strain of viral hemorrhagic septicemia virus (VHSV) associated with epizootic mortality in two new host species of Alaskan marine fish. Dis Aquat Organ 38:81–86

Meyers TR, Winton JR (1995) Viral hemorrhagic septicemia virus in North America. Ann Rev Fish Dis 5:3–24

Mortensen HF, Heuer OE, Lorenzen N, Otte L, Olesen NJ (1999) Isolation of viral haemorrhagic septicaemia virus (VHSV) from wild marine fish species in the Baltic Sea, Kattegat, Skagerrak and the North Sea. Virus Res 63:95–106

Morzunov SP, Winton JR, Nichol ST (1995) The complete genome structure and phylogenetic relationship of infectious hematopoietic necrosis virus. Virus Res 38:175–192

Mukherjee M, Chakrabarti A, Chakrabarty AN, Dastidar SG (1995) A pathogenicity model for the agents of epizootic ulcerative syndrome (EUS) in fish. Indian J Exp Biol 33:134–135

Nadin-Davis SA (2000) Rabies and rabies-related viruses. In: Thompson RCA (ed) Molecular epidemiology of infectious diseases. Oxford University Press, New York, pp 245–257

Neukirch M, Kunz U (2001) Isolation and preliminary characterization of several viruses from koi (*Cyprinus carpio*) suffering gill necrosis and mortality. Bull Eur Ass Fish Pathol 21:125–135

Nichol ST, Rowe JE, Fitch WM (1989) Glycoprotein evolution of vesicular stomatitis virus New Jersey. Virology 168:281–291

Nichol ST, Rowe JE, Winton JR (1995) Molecular epizootiology and evolution of the glycoprotein and non-virion protein genes of infectious hematopoietic necrosis virus, a fish rhabdovirus. Virus Res 38:159–173

Nishizawa T, Iida H, Takano R, Isshiki T, Nakajima K, Muroga K (2002) Genetic relatedness among Japanese, American and European isolates of viral hemorrhagic septicemia virus (VHSV) based on partial G and P genes. Dis Aquat Organ 48:143–148

Nishizawa T, Kurath G, Winton JR (1995) Nucleotide sequence of the 2 matrix protein genes (M1 and M2) of hirame rhabdovirus (HRV), a fish rhabdovirus. Vet Res 26:408–412

Nishizawa T, Yoshimizu M, Winton J, Ahne W, Kimura T (1991a) Characterization of structural protein of hirame rhabdovirus, HRV. Dis Aquat Organ 10:167–172

Nishizawa T, Yoshimizu M, Winton J, Kimura T (1991b) Comparison of genome size and synthesis of structural protein of hirame rhabdovirus, infectious hematopoietic necrosis virus, and viral hemorrhagic septicemia virus. Fish Pathol 26:77–78

Olesen NJ, Lorenzen N, Jorgensen PEV (1993) Serological differences among isolates of viral haemorrhagic septicaemia virus detected by neutralizing monoclonal and polyclonal antibodies. Dis Aquat Organ 16:163–170

Osadchaya YF, Nakonechnaya MG (1981) Rhabdovirus salmonis, the cause of a new disease in rainbow trout, *Salmo gairdneri*. J Ichthyol 21:113–121

Oshima KH, Arakawa CK, Higman KH, Landolt ML, Nichol ST, Winton JR (1995) The genetic diversity and epizootiology of infectious hematopoietic necrosis virus. Virus Res 35:123–141

Oshima KH, Higman KH, Arakawa CK, de Kinkelin P, Jorgensen PEV, Meyers TR, Winton JR (1993) Genetic comparison of viral haemorrhagic septicaemia virus isolates from North America and Europe. Dis Aquat Organ 17:73–80

Page RDM (1996) TREEVIEW: an application to display phylogenetic trees on personal computers. Comput Appl Biosci 12:357–358

Ristow SS, Arnzen JM (1989) Development of monoclonal antibodies that recognize a type-2 specific and a common epitope on nucleoprotein of infectious hematopoietic necrosis virus. J Aquat Anim Health 1:119–125

Rose JK (1980) Complete intergenic and flanking gene sequences from the genome of vesicular stomatitis virus. Cell 19:415–421

Rose JK, Gallione CJ (1981) Nucleotide sequences of the mRNA's encoding the vesicular stomatitis virus G and M proteins determined from cDNA clones containing the complete coding regions. J Virol 39:519–528

Ross K, McCarthy U, Huntly PJ, Wood BP, Stuart D, Rough EI, Smail DA, Bruno DW (1994) An outbreak of viral haemorrhagic septicaemia (VHS) in turbot (*Scophthalmus maximus*) in Scotland. Bull Eur Assoc Fish Pathol 14:213–214

Rowley H, Graham DA, Campbell S, Way K, Stone DM, Curran WL, Bryson DG (2001) Isolation and characterisation of rhabdovirus from wild common bream *Abramis brama*, roach *Rutilus rutilus*, farmed brown trout *Salmo trutta* and rainbow trout *Oncorhynchus mykiss* in Northern Ireland. Dis Aquat Organ 48:7–15

Roy P, Gupta KC, Kiuchi A (1984) Characterization of Spring viremia of carp virus mRNA species and the 3′ sequence of the viral RNA. Virus Res 1:189–202

Rucker RR, Whipple WJ, Parvin JR, Evans CA (1953) A contagious disease of salmon possibly of virus origin. U S Fish Wildl Serv Fish Bull 54:35–46

Sano T (1976) Viral diseases of cultured fishes in Japan. Fish Pathol 10:221–226

Sano T, Fukada H (1987) Principal microbial diseases of mariculture in Japan. Aquaculture 67:59–70

Sano T, Nishimura T, Okamoto N, Fukada H (1977a) Studies on viral diseases of Japanese fishes. VII. A rhabdovirus isolated from European eel *Anguilla anguilla.* Bull Jpn Soc Sci Fish 43:491–495

Sano T, Nishimura T, Okamoto N, Yamazaki T, Hanada H, Watanabe Y (1977b) Studies on viral diseases of Japanese fishes. VI. Infectious hematopoietic necrosis (IHN) in salmonids in the mainland of Japan. J Tokyo Uni Fisheries 63:81–85

Schäperclaus W (1938) Die Schädigung der deutschen Fischerei durch Fischparasiten und Fischkrankheiten. Fischerei-Ztg. 41:256–259, 267–270

Schäperclaus W (1990) Hämorrhagische Virus-Septikämie (HVS), Infektiöse Nierenschwellung und Leberdegeneration (INUL) In: Schäperclaus W, Kulow H, Schreckenbach K et al. (eds) Fischkrankheiten, Akademie-Verlag, Berlin, pp 297–310

Schlotfeldt HJ, Ahne W, Vestergard-Jorgensen PE, Glende W (1991) Occurrence of viral haemorrhagic septicaemia in turbot (*Scophthalmus maximus*) – a natural outbreak. Bull Eur Assoc Fish Pathol 11:105–107

Schütze H, Enzmann PJ, Kuchling R, Mundt E, Niemann H, Mettenleiter TC (1995) Complete genomic sequence of the fish rhabdovirus infectious haematopoietic necrosis virus. J Gen Virol 76:2519–2527

Schütze H, Enzmann PJ, Mundt E, Mettenleiter TC (1996) Identification of the non-virion (NV) protein of fish rhabdoviruses viral haemorrhagic septicaemia virus and infectious haematopoietic necrosis virus. J Gen Virol 77:1259–1263

Schütze H, Mundt E, Mettenleiter TC (1999) Complete genomic sequence of viral hemorrhagic septicemia virus, a fish rhabdovirus. Virus Genes 19:59–65

Smail DA (2000) Isolation and identification of Viral Haemorrhagic Septicaemia (VHS) viruses from cod *Gadus morhua* with the ulcus syndrome and from haddock *Melanogrammus aeglefinus* having skin haemorrhages in the North Sea. Dis Aquat Organ 41:231–235

Snow M, Cunningham CO (2000) Virulence and nucleotide sequence analysis of marine viral haemorrhagic septicaemia virus following in vivo passage in rainbow trout *Onchorhynchus mykiss*. Dis Aquat Organ 42:17–26

Snow M, Cunningham CO, Melvin WT, Kurath G (1999) Analysis of the nucleoprotein gene identifies distinct lineages of viral haemorrhagic septicaemia virus within the European marine environment. Virus Res 63:35–44

St Hilaire S, Ribble C, Traxler G, Davies T, Kent ML (2001) Evidence for a carrier state of infectious hematopoietic necrosis virus in Chinook salmon *Oncorhynchus tshawytscha*. Dis Aquat Organ 46:173–179

Stone DM, Ahne W, Denham KL, Dixon PF, Liu CT, Sheppard AM, Taylor GR, Way K (2003) Nucleotide sequence analysis of the glycoprotein gene of putative spring viraemia of carp virus and pike fry rhabdovirus isolates reveals four genogroups. Dis Aquat Organ 53:203–210

Stone DM, Way K, Dixon PF (1997) Nucleotide sequence of the glycoprotein gene of viral haemorrhagic septicaemia (VHS) viruses from different geographical areas: a link between VHS in farmed fish species and viruses isolated from North Sea cod (*Gadus morhua* L.). J Gen Virol 78:1319–1326

Strimmer K, von Haeseler A (1996) Quartet puzzling: a quartet maximum likelihood method for reconstructing tree topologies. Mol Biol Evol 13:964–969

Strimmer K, von Haeseler A (1997) Likelihood-mapping: a simple method to visualize phylogenetic content of a sequence alignment. Proc Natl Acad Sci U S A 94:6815–6819

Takano R, Mori K, Nishizawa T, Arimoto M, Muroga K (2001) Isolation of viruses from wild Japanese flounder *Paralichthys olivaceus*. Fish Pathol 36:153–160

Takano R, Nishizawa T, Arimoto M, Muroga K (2000) Isolation of viral haemorrhagic septicaemia virus (VHSV) from wild Japanese flounder *Paralichthys olivaceus*. Bull Eur Assoc Fish Pathol 20:186–192

Thiery R, de Boisseson C, Jeffroy J, Castric J, de Kinkelin P, Benmansour A (2002) Phylogenetic analysis of viral haemorrhagic septicaemia virus (VHSV) isolates from France (1971–1999). Dis Aquat Organ 52:29–37

Thiry M, Lecoq-Xhonneux F, Dheur I, Renard A, De Kinkelin P (1991) Sequence of a cDNA carrying the glycoprotein gene and part of the matrix protein M2 gene of viral haemorrhagic scepticaemia virus, a fish rhabdovirus. Biochim Biophys Acta 1090:345–347

Tordo N, Badrane H, Bourhy H, Sacramento D (1993) Molecular epidemiology of lyssaviruses: focus on the glycoprotein and pseudogenes. Onderstepoort J Vet Res 60:315–323

Tordo N, Poch O, Ermine A, Keith G, Rougeon F (1986) Walking along the rabies genome: is the large G-L intergenic region a remnant gene? Proc Natl Acad Sci U S A 83:3914–3918

Traxler GS, Roome JR, Kent ML (1997) Transmission of infectious hematopoietic necrosis virus in seawater. Dis Aquat Organ 28:31–38

Troyer RM, Kurath G (2003) Molecular epidemiology of infectious hematopoietic necrosis virus reveals complex virus traffic and evolution within southern Idaho aquaculture. Dis Aquat Organ 55:175–185

Troyer RM, LaPatra SE, Kurath G (2000) Genetic analyses reveal unusually high diversity of infectious haematopoietic necrosis virus in rainbow trout aquaculture. J Gen Virol 81:2823–2832

Vestergard-Jorgensen PE (1972) Egtved virus: antigenic variation in 76 virus isolates examined in neutralization tests and by means of fluorescent antibody technique. In: Mawdesley-Thomas LE (ed) Diseases of fish. Academic Press, London, pp 333–339

Walker PJ, Benmansour A, Calisher CH, Dietzgen R, Fang RX, Jackson AO, Kurath G, Leong JC, Nadin-Davies S, Tesh RB, Tordo N (2000) Family Rhabdoviridae. In: van Regenmortel MHV, Fauquet CM, Bishop DHL (eds) Virus taxonomy, classification and nomenclature of viruses. The seventh report of the international committee for taxonomy of viruses. Springer, Berlin Heidelberg New York, pp 563–583

Wang Y, Cowley JA, Walker PJ (1995) Adelaide River virus nucleoprotein gene: analysis of phylogenetic relationships of ephemeroviruses and other rhabdoviruses. J Gen Virol 76:995–999

Wang Y, Walker PJ (1993) Adelaide river rhabdovirus expresses consecutive glycoprotein genes as polycistronic mRNAs: new evidence of gene duplication as an evolutionary process. Virology 195:719–731

Wattanavijarn W, Wattanodorn S, Hunnak P, Tangtrongpiros J, Rattanaphani R (1986) Viruses of ulcerative diseased fish. Electron Microsc Soc Thailand Newsl 3:20–23

Wetzel T, Dietzgen RG, Geering AD, Dale JL (1994) Analysis of the nucleocapsid gene of lettuce necrotic yellows rhabdovirus. Virology 202:1054–1057

Wingfield WH, Fryer JL, Pilcher KS (1969) Properties of the sockeye salmon virus (Oregon strain). Proc Soc Exp Biol Med 130:1055–1059

Winter E, Yamamoto F, Amoguera C, Perucho M (1985) A method to detect and characterize point mutations in transcribed genes: amplification and overexpression of the mutant c-Ki-ras allele in human tumor cells. Proc Natl Acad Sci 82:7575–7579

Winton JR, Arakawa CK, Lannan CN, Fryer JL (1988) Neutralizing monoclonal antibodies recognize antigenic variants among isolates of infectious hematopoietic necrosis virus. Dis Aquat Organ 4:199–204

Winton JR, Batts WN, Nishizawa T, Stehr CM (1989) Characterization of the first North American isolates of viral hemorrhagic septicemia virus. Am Fish Soc Fish Health Sect Newl 17:2–3

Wolf K (1988) Fish viruses and fish viral diseases. Cornell University Press, Ithaca, pp 83–114

Wunner WH, Calisher CH, Dietzgen RG, Jackson AO, Kitajima EW, Lafon MF et al (1995) Family *Rhabdoviridae*. In: Virus taxonomy, classification and nomenclature of viruses, sixth report of the international committee for taxonomy of viruses. Springer, Berlin Heidelberg New York, pp 275–288

Yoshimizu M, Nishizawa T, Oseko N, Kimura T (1987) Rhabdoviral diseases of hirame (Japanese flounder). Fish Pathol 22:54–55

Zhang QY, Li ZQ, Gui JF (2000) Isolation of a lethal rhabdovirus from the cultured Chinese sucker *Myxocyprinus asiaticus*. Dis Aquat Organ 42:1–9

CTMI (2005) 292:119–141

Reverse Genetics on Fish Rhabdoviruses: Tools to Study the Pathogenesis of Fish Rhabdoviruses

M. Brémont

Unité de Virologie et Immunologie Moléculaires,
Institut National de la Recherche Agronomique, 78352 Jouy-en-Josas, France
michel.bremont@jouy.inra.fr

Abstract Rhabdoviruses, mainly in rainbow trout, are among the most devastating viruses for worldwide aquaculture. To date no effective treatments to fight against these viruses are available. During the past years, several approaches to develop efficient vaccines have been undertaken such as the use of immunogenic recombinant viral proteins, naked DNA or inactivated viruses. However, although these vaccines have been proven to be very effective on a small scale, they have never been used in the field because the vaccines would have to be injected into thousands of yearling trouts. The only alternative to injection consists of the development of attenuated live vaccines that can be administrated to trouts by bath immersion. Reverse genetics on trout rhabdoviruses offer the possibility of recovering a series of live recombinant viruses in which the viral genome has been irreversibly modified to generate cost-effective live, safe vaccines.

1 Introduction

Viruses belonging to the order of *Mononegavirale*, like the Rhabdoviruses, share the common structure of their genome, a single negative-strand RNA, and a similar replication strategy requiring a replicative complex consisting of the genomic RNA associated with three structural proteins, a nucleoprotein N, a large RNA-dependent RNA polymerase L, and a polymerase-associated component P. Based on this knowledge, some years ago Conzelmann's group developed a system for rabies virus (Schnell et al. 1994) allowing the recovery of infectious virus by mimicking the viral replication cycle in providing through transfection into cells virus-derived cDNAs encoding N, P, L and the genomic RNA. Since then, a number of negative-strand RNA viruses have been recovered by reverse genetics (for a review see Pekosz et al. 1999). However, all these viruses infect mammals and therefore have the ability to grow at high temperatures (>30°C). Here, rhabdoviruses infecting salmonids such as the infectious hematopoietic necrosis virus (IHNV) or the viral hemorrhagic septicemia virus (VHSV) represent a unique feature since for these viruses the optimal growth temperature is around 10–14°C. This particular characteristic probably made the development of reverse genetics for IHNV or VHSV more complex. IHNV and VHSV are both salmonid rhabdoviruses that are antigenically distinct with a genome structure similar to the mammalian rhabdovirus, with the exception of an additional NV gene located between the G and L genes. That NV gene encodes for a small nonstructural protein of unknown function (Kurath and Leong 1985; Basurco and Benmansour 1995). Fish rhabdoviruses possessing that NV gene have now been termed Novirhabdovirus. The entire nucleotide sequences of both IHNV and VHSV genomes have been determined (Morzunov et al. 1995; Schuetze et al. 1995, 1999).

Both IHNV and VHSV have a worldwide distribution and induce a very high mortality rate in infected fish farms. Infected fish generally die within 7–30 days after virus exposure. External signs of the virus-induced disease are an exophthalmia. Gills become pale and sometimes affected fish are darker than normal. Annual economic losses resulting from the viral infections in fish farms have been estimated to be more than US $23 million. These estimates indicate the real need for an efficient way to prevent these infectious diseases.

As there are no effective treatments against viruses in fish, the only method of controlling these diseases would be prevention by means of vaccination. For these viruses, strategies based on traditional attenuation of viruses or vaccines based on killed viruses have had only limited success (Lorenzen and Olesen 1997; Winton 1997). Vaccination trials have also been achieved using the recombinant G protein, since this glycoprotein is the viral protein responsible for immunogenicity and is the only protein that induces the production of neutralizing antibodies in infected animals (Lorenzen et al. 1990). For VHSV, however, administration of recombinant G protein induced low protection when expressed in prokaryotic systems (Lorenzen et al. 1993; Noonan et al. 1995), and moderate when expressed in insects (Lecocq-Xhonneux et al. 1994). Recently, DNA vaccines have been developed for VHSV and IHNV and have proven to be very effective in inducing protection (reviewed in Lorenzen et al. 2002). However, these DNA vaccines have to be administered by injection, and thus are less cost-effective and are restricted to medium-sized fish, and never yearlings, a stage in which the prevention of these diseases would be the most important. A breakthrough towards an effective protection of farmed fish against viral pathogens would be to obtain replicative live vaccines that may be used by bath immersion and thus are cost-effective.

In 1994, Schnell et al. (1994) succeeded in recovering recombinant rabies virus through the transfection into cells of plasmids containing virus-derived cDNAs. For the first time, a reverse genetics system for a negative-stranded RNA virus was established, roughly 15 years after the establishment of a reverse genetics system for a positive-stranded RNA virus, the poliovirus (Racaniello and Baltimore 1981). Negative-stranded RNA virus remained less amenable to artificial genetic manipulation for a long time, for at least three main reasons: (a), the large viral RNA dependent RNA polymerase has to be provided as a recombinant biologically protein; (b) the RNA genome to be replicated and encapsidated must have exact 3′ and 5′ ends; (c) genomic and antigenomic RNA are never naked inside the cell but always associated in a ribonucleoprotein (RNP) complex consisting of N, P and L. All these obstacles have been progressively surmounted so that at the end the recombinant negative-stranded RNA viruses could be recovered from cDNA. For fish rhabdoviruses, similar strategies have been developed, but one must bear in mind that the growth temperature of these viruses was drastically different from the mammalian rhabdovirus such as VSV or rabies virus (14°C vs. 37°C). That difference represented an additional obstacle for the establishment of a reverse genetics system for fish rhabdoviruses.

2
Minigenome Constructs for Salmonid Rhabdoviruses

For negative-stranded RNA viruses, a first approach towards the recovery of recombinant infectious virus has been to engineer a model genome called a minigenome, which is the viral RNA genome in which all the coding regions have been deleted and replaced by a reporter gene such as the chloramphenicol acetyl transferase (CAT), luciferase (LUC), or the green fluorescent protein (GFP). This minigenome is a unique tool that makes it possible to establish all the parameters required to overcome all the obstacles mentioned above. The first demonstration of this approach was provided by Krystal and his colleagues, who developed a Sendaï virus-based minigenome (Park KH et al. 1991).

Thus, as a first step towards establishing a reverse genetics system for salmonid rhabdoviruses, an IHNV-derived cDNA plasmid construct, pIHNV-CAT(-), in which all the IHNV coding regions were deleted and replaced by the chloramphenicol acetyltransferase (CAT) reporter gene in negative sense, was engineered. As shown in Fig. 1A, a T7 promoter sequence and a hepatitis δ virus antigenome ribozyme sequence (Perrota and Been 1991) were fused to the IHNV trailer/L-gene end and leader/N-gene start sequences, respectively. The use of a hepatitis δ virus antigenome ribozyme sequence allowed the generation of an RNA minigenome ending at the authentic extreme nucleotide of the 3′ end IHNV genomic RNA, through the autocatalytic activity of the ribozyme. The use of a ribozyme sequence to generate the exact terminus of an RNA genome was first reported for Nodavirus and VSV RNAs (Ball 1992; Pattnaik et al. 1992). Through in vitro RNA polymerase-driven transcription of the pIHN-CAT(–) plasmid construct, analysis of the generated RNA demonstrated the efficient cleavage of the ribozyme sequence (data not shown).

To demonstrate that this IHNV-derived RNA minigenome could be encapsidated and replicated by the wild-type IHNV, IHNV-infected *epithelioma papulosum cyprini* (EPC) cell monolayers (Fijan et al. 1983) were transfected with the IHNV-CAT (–) RNA following the strategy depicted in Fig. 1B.

After the first and second passages, the supernatant was collected for CAT activity by ascending thin-layer chromatography (TLC) of the cell lysates following incubation with [^{14}C]chloramphenicol as substrate (Gorman et al. 1982). As shown in Fig. 1C, CAT activity, although at a low level, was detectable after both passages one and two.

CAT activity observed after the first passage demonstrated that the RNA minigenome had been encapsidated by the IHNV proteins provided in trans by IHNV infection and CAT activity after the second passage demonstrated the replication of the encapsidated minigenome. It is notable that this rescue of exogenous RNA by a helper virus has been described for a number of single-stranded negative-sense RNA viruses belonging to the family *Paramyxoviridae* (Park et al. 1991; Collins et al. 1991; Yunus et al. 1999; Sidhu et al. 1995; De Banerjee 1993; Dimock and Collins 1993; Randhawa et al. 1997). A similar rescue experiment has been described more recently for VHSV, another *Novirhabdovirus* (Betts and Stone 2001). However, it has never been described for other members of the *Rhabdoviridae* family such as vesicular stomatitis virus or rabies virus (Conzelmann 1998).

The ability to rescue the T7-driven in vitro-synthesized RNA minigenome prompted us to investigate whether the T7 RNA polymerase could be provided in fish cells by infection with the recombinant vaccinia virus vTF7-3

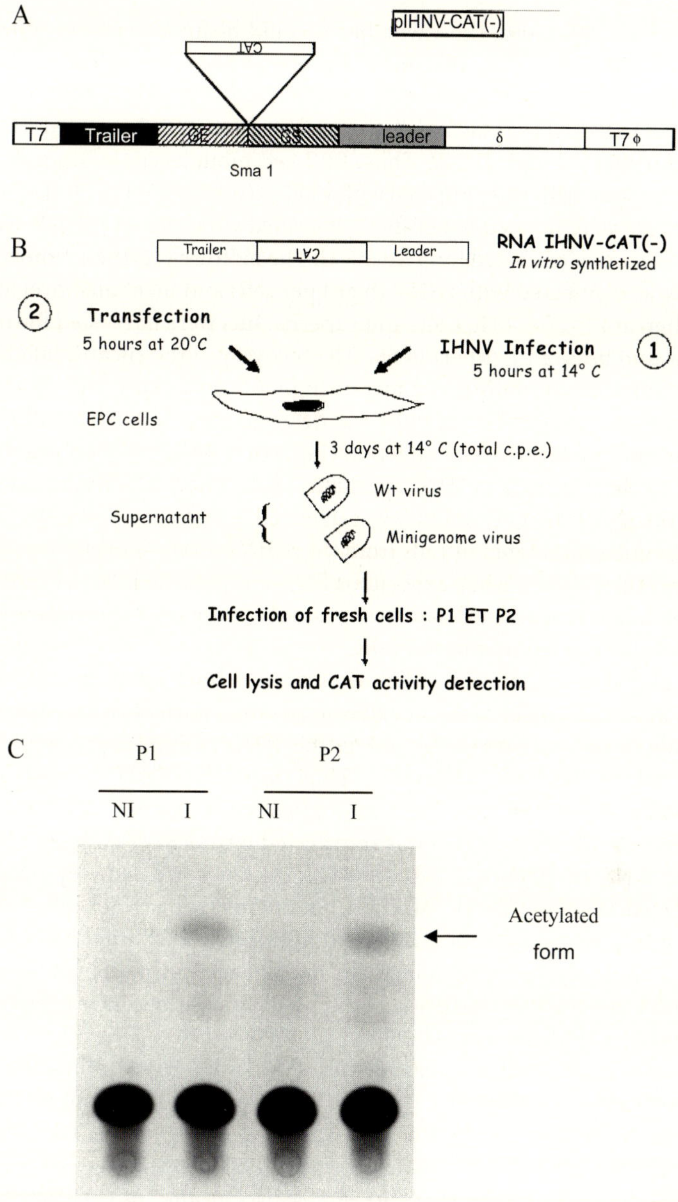

Fig. 1A–C Expression of salmonid rhabdovirus minigenome by RNA transfection. Minigenome pIHNV-CAT (–) (**A**) was used to make RNA transcripts in vitro and then transfected into IHNV-infected EPC using the strategy (**B**). The CAT activity in the EPC cells was assayed using the TLC method (**C**). *P1*, the first passage; *P2*, the second passage; *NI*, not infected; *I*, infected

(Fuerst et al. 1986) and, thus, whether it could be an alternative system for rescue experiments. The experimental procedure is schematically presented in Fig. 2A.

The pIHNV-CAT(–) plasmid was transfected in vTF7-3-infected EPC cells and superinfected with IHNV. Thus, EPC cell monolayers in six-well plates (3×10^4 cells per well) were infected with vTF7-3 (m.o.i. of 5) for 1 h at 37°C. Cell monolayers were washed twice and transfected with 1 μg of pIHNV-CAT(–). These cells were incubated for 5 h at 37°C. The mixture was then removed and the cells were infected with IHNV (5 pfu per cell) and incubated overnight at 20°C, then at 14°C for 48 h. Cells and supernatants were harvested for further analysis and passaging experiments. The recovery of the encapsidated IHNV minigenome was monitored by measuring CAT gene expression in infected cells after one passage. CAT activity was detected as early as 8 h after infection and was optimal at 30 h after infection (Fig. 2B). CAT expression was directly related to the presence of IHNV proteins, since when IHNV superinfection was omitted (NI) no CAT activity was detected. The rescue of the IHNV RNA minigenome synthesized in cells from the pIHNV-CAT(–) plasmid proved to be more efficient than when exogenous RNA was provided. The encapsidated minigenome was replicated after two passages of the cell supernatant as CAT activity was detected in infected cells.

Although the leader and trailer sequences of both IHNV and VHSV are largely different, 10 out of the 12 extreme terminal nucleotides are preserved, and thus it was of interest to rescue the IHNV minigenome with VHSV as the helper virus and vice-versa. Thus, a pVHSV-CAT(–) construct derived from the VHSV genome was engineered as for the pIHNV-CAT(–) minigenome. pVHSV-CAT(–) was transfected into vTF7-3-infected EPC cells and the RNA minigenome was shown to be encapsidated, replicated and propagated following VHSV infection, but not following IHNV infection. The IHNV minigenome was not propagated following VHSV infection (data not shown). These results together indicate that other *cis*-acting elements in IHNV and VHSV genomes are not preserved. These observations have been recently confirmed by Hoffmann et al. (2003).

We also engineered a construct (Fig. 3A) of an IHNV miniantigenome (positive sense), pIHNV-CAT (+). The positive-sense minigenome was rescued the same way as was the negative-sense minigenome. At 24 h after transfection and virus infections (IHNV and vTF7-3), cells were lysed and analyzed for CAT activity (passage P0). A background of CAT activity was observed (Fig. 3B, left lane) when cells were infected only with vTF7-3 and transfected with pIHNV-CAT(+), due to the messenger sense of RNA synthesized by the T7 RNA polymerase. However, when cells were superinfected with IHNV, CAT activity was dramatically increased, demonstrating that the

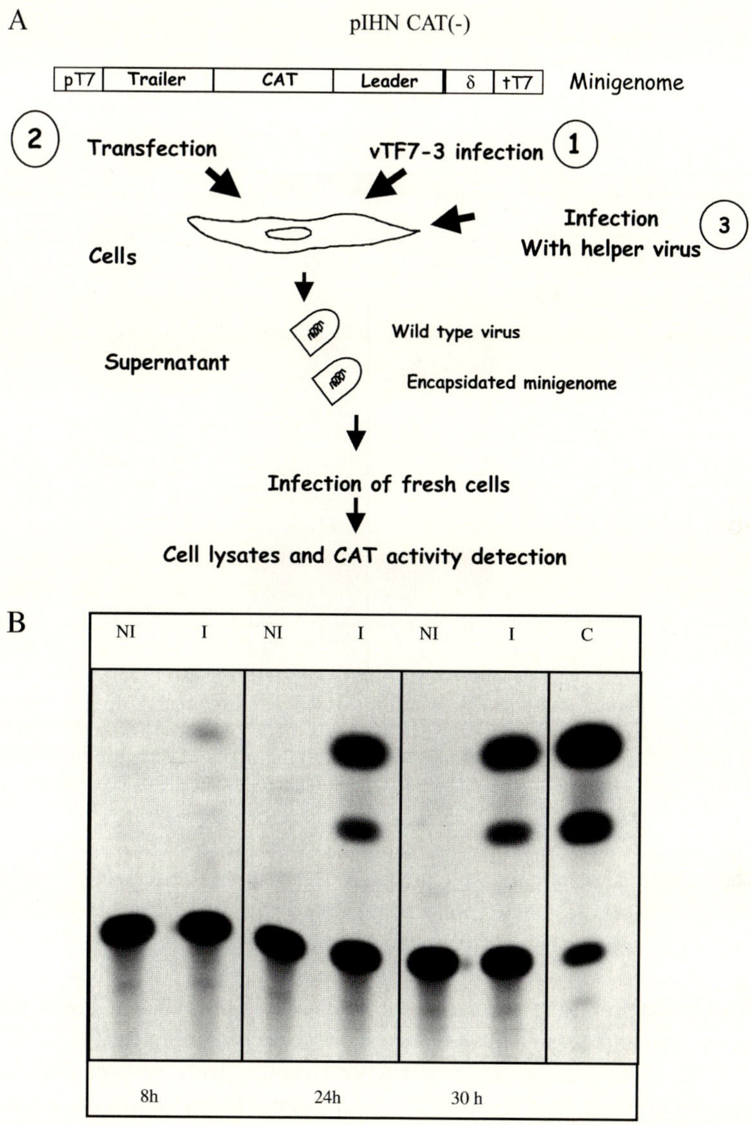

Fig. 2A, B Expression of salmonid rhabdovirus minigenome by DNA transfection. Minigenome pIHNV-CAT (–) was directly transfected into vTF7-3 and IHNV-infected EPC using the strategy (**A**) and the CAT activity in the EPC cells was assayed using the TLC method at 8, 24, and 30 h after transfection (**B**). *NI*, not infected; *I*, infected. *C*, positive control (plasmid encoding CAT gene)

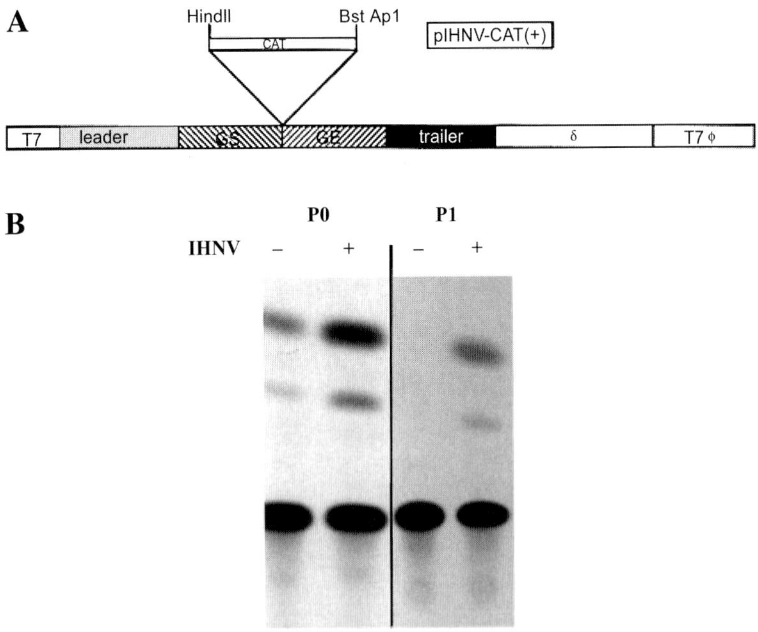

Fig. 3A, B Expression of the positive-strand salmonid rhabdovirus minigenome. The positive-strand minigenome pIHNV-CAT (+) was made (**A**) and transfected into EPC cells. The CAT activity in the EPC cells was assayed using the TLC method with (+) or without (–) infection with IHNV (**B**). *P0*, transfected cells; *P1*, the first passage

miniantigenome construct replicated. CAT activity was detected at P1 as well, indicating that the construct was encapsidated and replicated.

For mammalian rhabdoviruses such as rabies virus and vesicular stomatitis virus, it has been shown that minigenomes are rescued when the viral nucleocapsid proteins are expressed from cDNAs in cells (Pattnaik et al. 1992; Conzelmann and Schnell 1994).Thus, sequences containing IHNV genes encoding N, P, L, and the Nv nonstructural protein were recovered by PCR from a full length DNA copy of the IHNV genome using specific primers. PCR products were inserted into the pET-14b vector (Novagen), resulting in plasmids pT7-N, pT7-P, pT7-Nv, and pT7-L. In these constructs, the IHNV coding regions were inserted between the T7 promoter and the T7 terminator sequences. A test was developed to evaluate whether the three IHNV-derived replicative complex proteins were expressed and functionally active in EPC cells after transfection of the respective recombinant pT7-N, pT7-P and pT7-L plasmids (Fig. 4A).

A

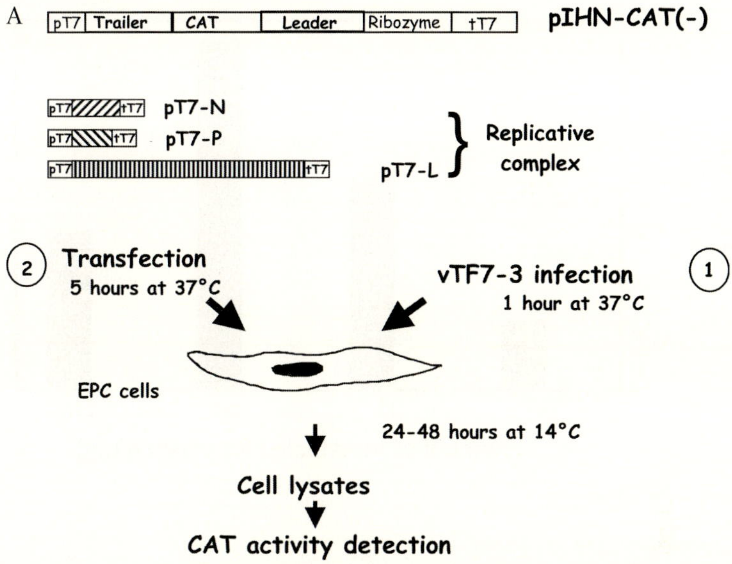

B

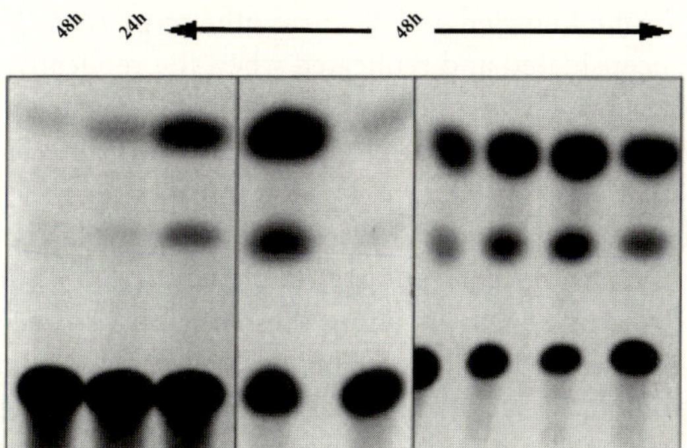

pIHNV	1	1	0.5	0.5	0.5	0.5	0.5	0.5
pT7-N	2.5	2.5	0.5	1	0.5	0.5	0.5	0.5
pT7-P	1.2	1.2	0.2	0.2	0.2	0.2	0.2	0.2
pT7-L	0.0	1.2	0.2	0.2	0.2	0.2	0.2	0.2
pT7-Nv	0.0	0.0	0.0	0.0	0.0	0.05	0.125	0.25
pBS	0.0	0.0	0.5	0.0	0.25	0.20	0.125	0.0

Fig. 4A–C Expression of salmonid rhabdovirus minigenome by DNA transfection in cells expressing IHNV proteins. Minigenome pIHNV-CAT (–) was directly transfected into vTF7-3-infected EPC using the strategy (**A**). Instead of using IHNV superinfection, plasmids expressing IHNVN, P, L and NV were co-transfected into EPC cells. The CAT activity in the EPC cells was assayed using the TLC method when different concentrations of these plasmids were used (**B**). CAT activity was quantitated using phosphoimaging when different concentrations of pT7-Nv were used (**C**) (*continued on next page*)

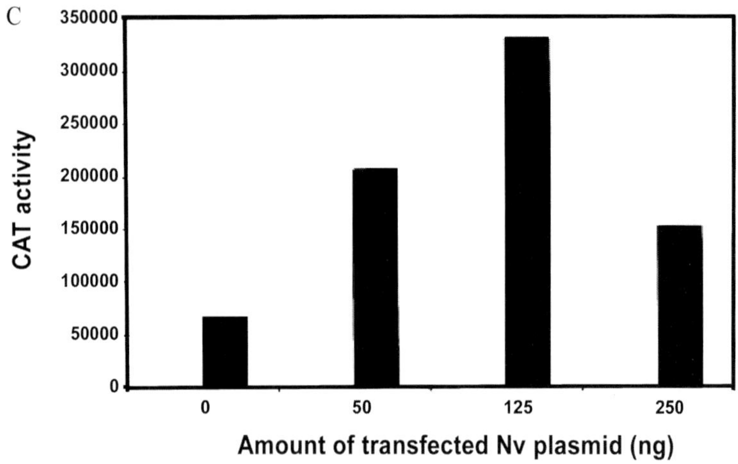

Fig. 4A–C (*continued*)

Based on the knowledge that the synthetic pIHNV-CAT(–) minigenome could be encapsidated and replicated when the replicative complex proteins were provided by IHNV infection, minigenome constructs were transfected into vTF7-3-infected EPC cells together with two different combinations of pT7-N, pT7-P and pT7-L. CAT activity was monitored at 24 and 48 h after transfection. Depending on the ratio of the four plasmid constructs used, CAT activity was variable but detectable in all cases (Fig. 4B), indicating that the three constructs encoding the replicative complex were functional and able to encapsidate, transcribe, and replicate the IHNV minigenome. The addition of pT7-N plasmid in excess drastically reduced the level of CAT activity, contrasting with the results obtained with respiratory syncytial virus, a paramyxovirus, for which replication of a synthetic minigenome increased after the addition of a plasmid encoding the nucleoprotein (Fearns et al. 1997).

IHNV and VHSV are both rhabdoviruses that replicate at 14°C in fish cells and encode a nonviral (NV) protein. They belong to a genus called *Novirhabdovirus*. Using the optimal conditions determined above, the effect of adding increasing amounts of the pT7-Nv construct was determined. Quantification of CAT activity was performed by phosphorimaging of the acetylated spots detected on TLC plates. As shown in Fig. 4C, a roughly fivefold increase in CAT activity was observed when the pT7-Nv expression plasmid was added to the transfection mixture in catalytic amounts (0±125 μg). Thus, the Nv protein may play a role at either replication or the transcription step, at least in this minigenome system. This result contrasts with those recently published for another novirhabdovirus, snakehead rhabdovirus (SHRV), which replicates at a high temperature of 31°C (Johnson et al. 2000), and for which a reverse genetics system allowing the recovery of infectious virus entirely from cDNAs has been established. In the SHRV system, when the Nv gene is mutated to introduce a premature stop codon in the coding region, no differences in the virus titer are observed compared to the wild-type SHRV.

The use of minigenome constructs expressing a reporter gene established the conditions for the transfection and vTF7-3 infection of fish cells, and demonstrated that biologically active IHNV proteins could be provided in trans from IHNV-derived plasmid constructs.

2.1
Recovery of IHNV from cDNAs

The ability to generate infectious virus derived from cDNA is a very powerful approach, since the viral genome can be manipulated to generate attenuated vaccine strains and also makes the introduction of genetic tags feasible, thus discriminating between field and the vaccine strains. Moreover, an extra gene can be stably introduced into the viral genome, and thus negative-strand RNA viruses such as IHNV could be used as a gene vector (Conzelmann 1998; Pekosz et al. 1999). To evaluate the feasibility of generating a live attenuated viral vaccine strain to prevent IHNV infection in the field, a reverse genetic system for IHNV allowing recovery of genetically tagged infectious virus through cDNA transfection into fish cells has been developed.

A full-length DNA copy of the IHNV RNA genome has been assembled from overlapping IHNV-derived cDNA following the strategy depicted in Fig. 5. Four overlapping cDNA fragments (numbered 1 to 4) covering the full-length IHNV RNA genome were generated by RT-PCR with specific primers, and assembled in a pBlueScript plasmid backbone. Restriction enzyme sites used for the construct were all present in the original sequence (Schutze et al. 1995), with the exception of the *PstI* site. The pTϕRibo plasmid that served for the insertion of fragment 4 had been previously described (Biacchesi et al. 2000b).

The final construct, termed pIHNV-Pst, encodes an antigenomic (positive sense) IHNV RNA starting and ending at the exact authentic 5′ and 3′ nucleotides. Using the conditions established for the rescue of IHNV minigenome, vTF7-3-infected EPC cells were transfected with the pIHNV-Pst construct, together with the pT7-N, pT7-P and pT7-L plasmids. Recovery of recombinant IHNV was successful, although at very low titer ($<10^3$ pfu/ml). After additional passages of the viral supernatant onto fresh cells, total RNA

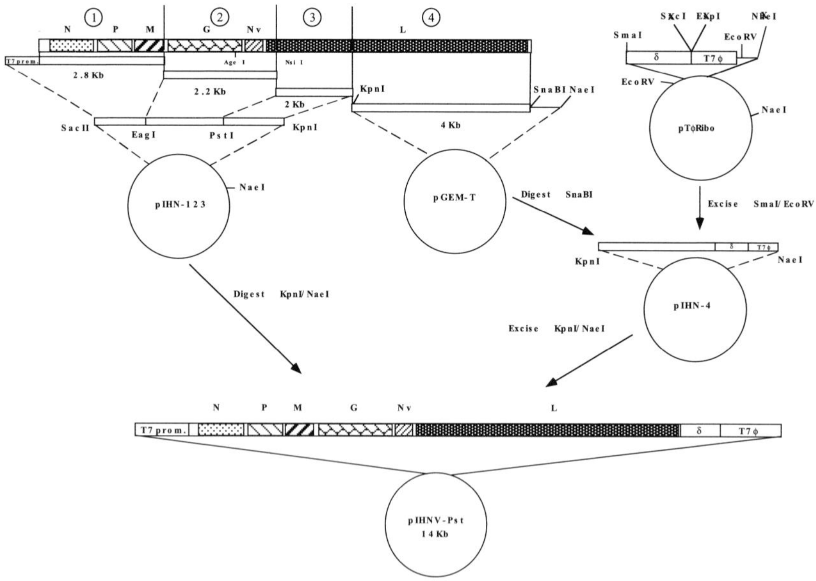

Fig. 5 Construction of the full-length infectious clone. Four overlapping fragments covering the full-length IHNV RNA genome were generated and ligated into the pBluescript plasmid backbone using restriction sites indicated. *T7prom*, T7 RNA polymerase transcription promoter; *T7φ*, T7 RNA polymerase transcription terminator; *δ*, antigenome hepatitis delta virus ribozyme sequence

of infected cells were extracted and used for RT-PCR experiments using specific primers to confirm the recovery of recombinant IHNV by nucleotide sequencing (Fig. 6).

2.2 Targeted Modifications of the IHNV Genome: Exchange of VHSV and IHNV Proteins

As the final goal of establishing a reverse genetics system for IHNV is the development of attenuated live vaccine administrable through bath immersion, it is important to demonstrate that the IHNV genome can be manipulated. The first approach was to see whether the gene encoding the IHNV G could be replaced by other heterologous G genes. Thus, the IHNV genome was engineered such that unique restriction enzyme sites (*SpeI* and *SmaI*) were added at the initiation and the stop codon of the G open reading frame. To replace the IHNV G gene, G genes from the VHSV and the spring viremia of carp

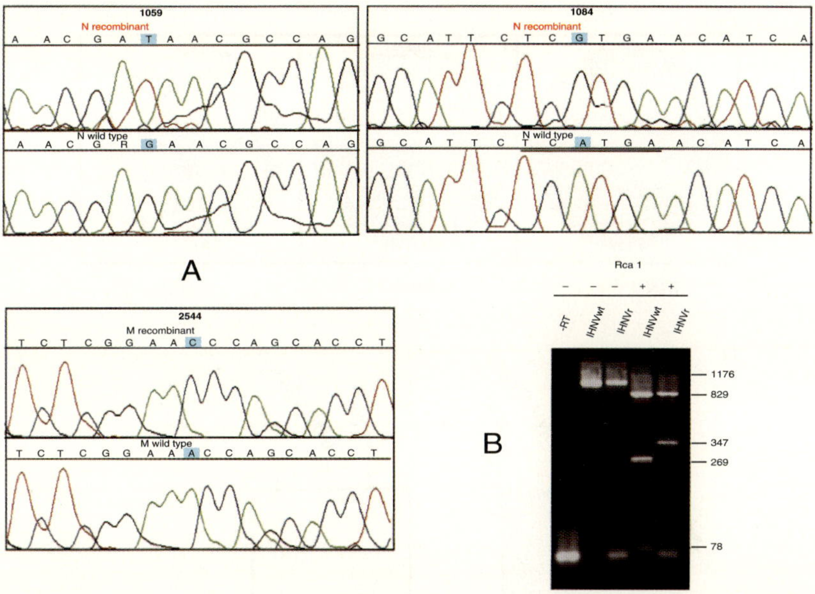

Fig. 6A, B Genetic tags in rIHNV. Electropherograms showing nucleotide sequence of part of the N and M PCR products generated from the wtIHNV and rIHNV genomes were sequenced across each of the three selected tags (*blue boxes*). The RcaI restriction site in the wt sequence is *underlined* (**A**). Gel analysis of the RT-PCR product used to identify rIHNV (**B**). Total RNA extracted from cells infected with wtIHNV or rIHNV was used to amplify the N gene by RT-PCR with N-specific primers. PCR products were analyzed on 1% agarose gel with (+) or without (–) restriction enzyme digestion. A control without RT (*–RT*) was added to ensure that no contaminating plasmids were present in rIHNV-infected cell supernatants. The sizes of the fragments in nucleotides are indicated on the *right*

virus (SVCV), a carp rhabdovirus belonging to the vesiculoviruses, were amplified through RT-PCR on viral genomic RNAs and inserted into the pIHNV digested with *SpeI* and *SmaI* . Corresponding chimeric recombinant viruses, rIHNV-G_{vhsv} and rIHNV-G_{svcv} were generated and produced. Growth of both recombinant viruses in cell culture was shown to be similar to the wild-type rIHNV. For the rIHNV-G_{vhsv} the only remarkable feature was the size of the induced plaques, which were of intermediate size between induced plaques in VHSV-infected and IHNV-infected cells (Fig. 7A). Direct confirmation of the recovery of recombinant IHNV expressing the VHSV G instead of IHNV G was provided by analysis of gradient-sucrose purified viruses on SDS-PAGE and Western blot assay using monoclonal antibodies directed against the IHNV P and the VHSV G (Fig. 7B).

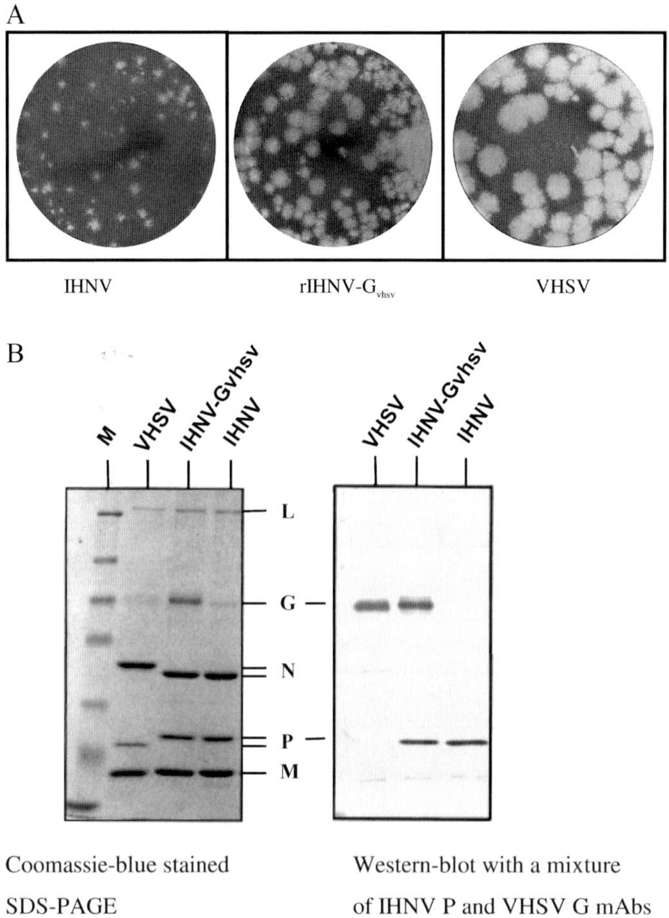

Fig. 7A–C Replacing the G of IHNV with that from VHSV or SVCV. rIHNV with the G derived from VHSV formed plaques sized between the IHNV and VHSV (**A**). Proteins extracted from cells infected with IHNV, VHSV and the recombinant rIHNV-Gvhsv were subjected to SDS-PAGE and stained with Coomassie-blue staining (**B**, *left*) or Western-blotted with anti-P or anti-G mAbs (**B**, *right*). Proteins extracted from cells infected with IHNV, SVCV and the recombinant rIHNV-Gsvcv were subjected to SDS-PAGE and stained with Coomassie-blue staining (**C**)

Demonstration was also provided that a chimeric IHNV expressing a glycoprotein from a vesiculovirus such as the spring viremia of carp virus (SVCV) was possible (Fig. 7C). Together, these data show that the IHNV glycoprotein can be efficiently replaced with that of either of two unrelated fish rhabdoviruses, VHSV (*Novirhabdovirus*) and SVCV (*Vesiculovirus*). The observa-

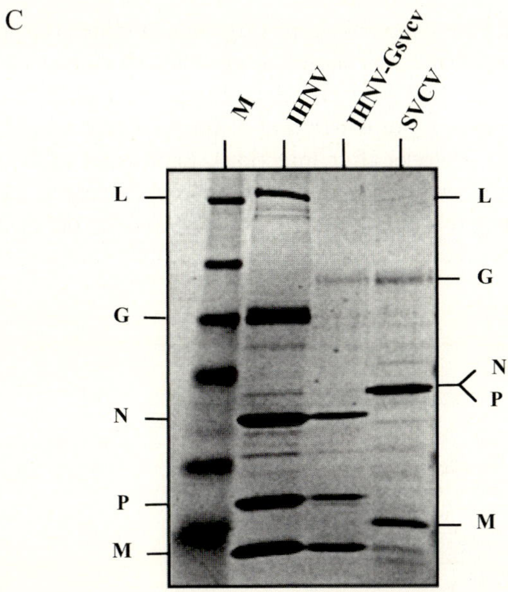

Fig. 7A–C (continued)

tion that this process does not require specific IHNV G-derived sequences fits well with the results from previous studies on VSV (Kahn et al. 1999; Kretzschmar et al. 1997; Schnell et al. 1996) but is not in good accordance with the results obtained with RV (Mebatsion et al. 1996, 1997), although IHNV is more related to lyssaviruses such as RV than to vesiculoviruses such as VSV.

In addition to the exchange of IHNV glycoprotein by heterologous glycoproteins, the IHNV matrix M protein could also been exchanged by that of VHSV. The growth of these two recombinant IHNVs was strongly reduced in cell culture, indicating that some steps in virus assembly were strongly affected (Biacchesi et al. 2002). Taken together, these data show the extreme flexibility of IHNV with regard to the functional replacement of two major structural proteins, M and G.

3
IHNV as Gene Vector

The IHNV genome has been modified by adding an expression cassette. A unique restriction enzyme site located in the intergenic region between the M and G gene has been used to insert a new cistron encoding for the green fluorescent protein (GFP) as a reporter gene (Fig. 8A). Modified recombinant IHNV has been recovered and was shown to grow in cell culture as well as the wild-type virus (titer >10^8 pfu/ml). In addition, induced mortality by the rIHNV-GFP in trouts was similar to that of the wild type (roughly 90% of cumulative mortality 2 weeks after Infection). Expression of the GFP in rIHNV-GFP infected cells was detectable at a high level as early as 12 h after infection, demonstrating that the cassette was functional for the expression

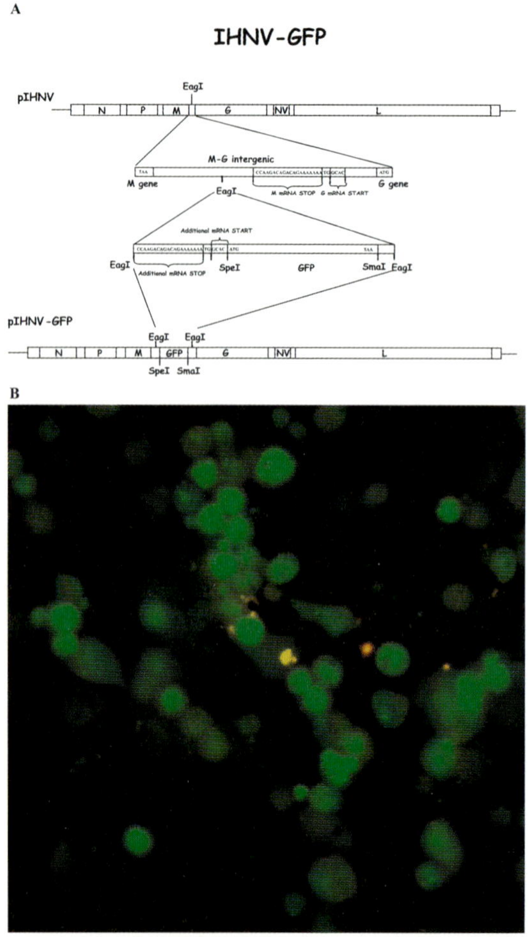

Fig. 8A, B Construction of rIHNV-GFP and expression of GFP in cells infected with the recombinant virus. **A** GFP was cloned into rIHNV in the intergenic sequences between the M and G. **B** Cells infected with rIHNV-GFP was used to infect EPC cells and observed under a fluorescent microscope

of additional genes in the IHNV genome (Fig. 8B). Expression of GFP in EPC-infected cells and in infected fish remained stable for several passages, demonstrating that IHNV could potentially be used in the future as a gene vector for the delivery of genes of interest in fish.

4
Role of NV in Virus Replication

The NV gene is present, so far, only in some fish rhabdoviruses grouped in the novirhabdovirus family, and proteins encoded by NV genes are only distantly related. The role and putative function of this nonviral protein is unclear. Reverse genetics on IHNV allowed investigating whether (a) NV is required for viral replication and (b) NV plays a role in fish. Thus a recombinant IHNV was engineered such that the NV gene was deleted and replaced by the GFP gene (rIHNV-ΔNV-GFP). In cell culture, the rIHNV-ΔNV-GFP was shown to be strongly affected in the replication since following successive passages in cell culture, the viral titer progressively decreased and was lower than 10^3 pfu/ml after three or four passages. Interestingly, the NV from VHSV could functionally replace the IHNV NV since a recombinant IHNV expressing the VHSV NV gene instead of its own NV gene (rIHNV-NV_{VHSV}) replicated as well as the wild-type IHNV in cell culture (Thoulouze et al. 2004). Thus, it can be assumed that NV is required for efficient replication of novirhabdoviruses

```
SNKH      1 ...MPG.NNHNDRPVAL...IMSRLSRDPRDCIHHTVDTRGMTPGKIIHQVIPPHPTQMR
VHSV      1 MATQPALSTTSFSPLVLREMITHRLKFDPSNYLNCDLDRSDISPVDFFETTLPRILDDLR
IHNV      1 ....MDHRDINTNMEALREVLRYKNKVAGHGFLFDDGDLVWREEDDATWRRLCDVVNALI
HIRA      1 ....MNSKTPSTDIAALKDLLRYKVTVARHGFLFDDGKIVWSEDGDEAWNRLLVVVGALR

SNKH     54 HTNRPSPPVILNLQI.LSEGGHVCARDVDTADPVP.RAGPERTIQWGTDRDATEELASPA
VHSV     61 ASTRLPHLHVLDMRISLLERTHYMFRNVPSSPATTGRLTDPELVLISHAEVGLLTRGSGL
IHNV     57 SSKRMQRVLYMDLSITKGEG.HLLFVDLQGTKNRLHKEPRFRRHLILIEDFLAYPR....
HIRA     57 SSNRMSQALFMDMSITKGDG.YLLFSDLQGTNNLQYRTPKFRQYLFPVDEFLPLPR....

SNKH    112 PEEIYEDNDSW
VHSV    121 PS.........
IHNV        ...........
HIRA        ...........
```

Fig. 9 Comparison of the amino acid NV sequences among the fish rhabdoviruses. *SNKH*, Snakehead rhabdovirus; *VHSV*, viral hemorrhagic septicemia virus; *IHNV*, Infectious hematopoietic necrosis virus; *HIRA*, Hirame rhabdovirus

in cell culture and that despite a low percentage of amino acid identity between the IHNV and VHSV NV proteins (Fig. 9), both proteins may play a similar role in cell culture (Thoulouze et al. 2004).

4.1 Role of the NV Protein in IHNV Pathogenesis in Fish

All the viruses described in the present report were compared for their ability to replicate in vivo and to induce symptoms of disease in trout. Samples of 75 juvenile trouts (mean weight, 0.4 g) were infected by bath immersion with 5.10^4 pfu/ml of the wt rIHNV, rIHNV-ΔNV-GFP and rIHNV-NV_{vhsv}. As positive control, the wt IHNV 32/87 strain was included. Mortalities were recorded every day for 2 weeks following virus exposure. As shown in the Fig. 10, wt IHNV 32/87, wt rIHNV and rIHNV-NV_{vhsv} are all highly pathogenic for trouts since cumulative mortalities ranged between 50% and 70% 2 weeks after infection. Trouts infected with these viruses developed typical symptoms of IHNV infection and died between days 5 and 6 after exposure. In contrast,

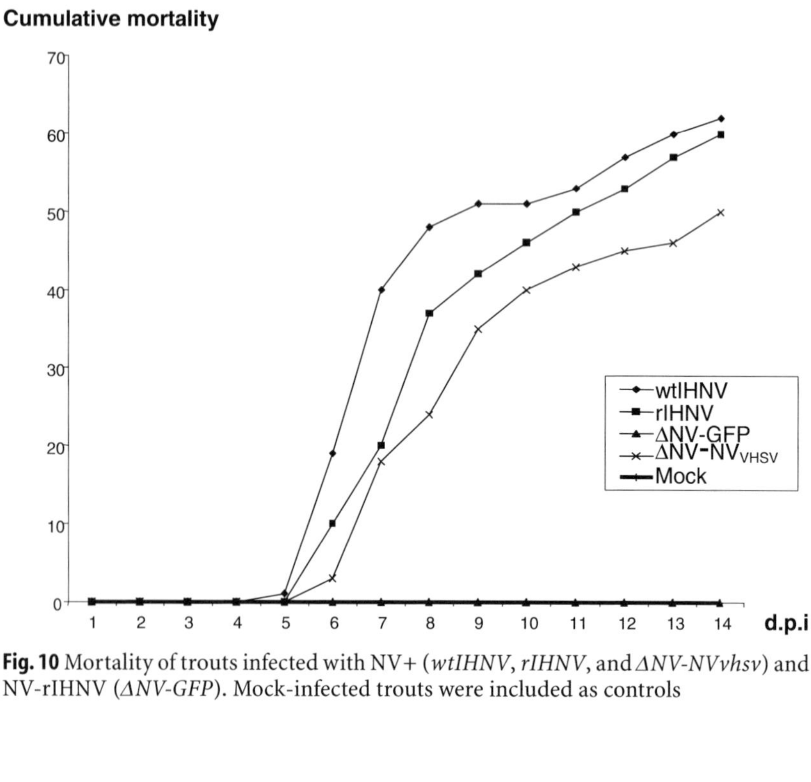

Fig. 10 Mortality of trouts infected with NV+ (*wtIHNV*, *rIHNV*, and *ΔNV-NVvhsv*) and NV-rIHNV (*ΔNV-GFP*). Mock-infected trouts were included as controls

trouts infected with the rIHNV-ΔNV-GFP did not develop any external signs of disease and no deaths were recorded. In every case, virus was recovered from sampled survivors or dead fish. Thus, these experiments show that NV gene is essential for IHNV pathogenicity in trouts and that NV_{VHSV} protein can functionally replace the NV_{IHNV} protein both in vitro and in vivo.

To rule out the possibility that the absence of mortality observed when trouts were exposed to the rIHNV-ΔNV-GFP could be due to the means of administration, experimental infections were also carried out by injection and deaths were recorded every day. Similar results as those recorded by bath immersion infection were obtained following virus injection. Thus, differences in virus pathogenicity were not to the result of differences in virus entry into the trout and were independent of the means of administration. These results were consistent with a loss of pathogenicity due to NV deletion, which is reverted by NV_{VHSV} expression. Together the data demonstrated that both NV proteins from IHNV and VHSV play an essential and similar role for the pathogenicity of novirhabdovirus in trouts.

5 Conclusion

In this review, we have shown that infectious IHNV was successfully recovered from fish cells following infection with the vTF7-3 recombinant vaccinia virus and cotransfection with T7-driven expression plasmids containing a positive-sense copy of the IHNV genome and the three genes encoding the nucleocapsid proteins N, P, and L, following a strategy initially developed for mammalian rhabdoviruses, rabies virus and VSV (Lawson et al. 1995; Schnell et al. 1994; Whelan et al. 1995). The main obstacle to establish a reverse genetic system for IHNV, based on the use of vTF7-3, was to find temperature conditions compatible with vTF7-3 replication (30°C) and IHNV growth (14°C). This problem was overcome when fish cells were infected with vTF7-3 and transfected with the expression plasmids at 37°C for 7 h before shifting the temperature to 14°C. Under these conditions, we observed that during the few hours at 37°C, the vTF7-3-infected cells expressed the majority of T7 RNA polymerase, allowing the transcription of the transfected expression plasmids. Moreover, since vTF7-3 does not replicate in fish cells at 14°C (Biacchesi et al. 2000; Johnson et al. 2000), the cells could be kept for at least 1 week after infection and after transfection before passage of the viral supernatant, to optimize the recovery of rIHNV. The rIHNV was morphologically indistinguishable from the wild-type virus by electron microscopy observations (data not shown) and rIHNV replicates in cell culture and was pathogenic in juvenile rainbow trout such as the wtIHNV. The addition of a plasmid encoding the nonstructural NV protein to the transfection mix slightly increased the efficiency of IHNV recovery in terms of the number of infected-cell foci after one passage of the viral supernatant. However, although the role of the NV protein in IHNV replication in cell culture is not yet clear, it has been demonstrated that NV was necessary for an efficient replication of the virus in cell culture.

The extreme flexibility of IHNV with regard to the functional replacement of two major structural proteins, M and G, with those of a distantly related fish rhabdovirus have been demonstrated (Biacchesi et al. 2002). For example, a recombinant virus, rIHNV-Gvhsv, in which the IHNV G gene was replaced with that of VHSV, another antigenically distinct novirhabdovirus, was shown to grow as well as wild-type IHNV and VHSV in cell culture (reaching a titer of 4 $X10^8$ pfu/ml), and to be as pathogenic as wild-type viruses in fish. Furthermore, a recombinant virus, rIHNV-Gsvcv, was recovered, demonstrating that a glycoprotein originating from another viral genus can be efficiently incorporated into IHNV virions. Finally, the role of the NV protein for the pathogenicity of IHNV in rainbow trout has been also demonstrated, since recombinant IHNV deleted for the NV gene were totally attenuated in rainbow trouts by both bath immersion and injection (Thoulouze et al. 2004). These observations now enable the future development of new viral strains that can be used as attenuated live vaccines administrable by bath immersion.

Acknowledgements Work presented in this review was carried out with financial support from the Commission of the European Communities, Agriculture and Fisheries (FAIR) specific RDT program, CT98–4398. I thank all the members of the Fish Molecular Virology Group (INRA, Jouy-en-Josas, France) who contributed to this work and more specifically Dr. Stéphane Biacchesi, Dr. Maria Isabel Thoulouze and Monique LeBerre. Dr Michel Dorson and the fish facility staff (INRA, Jouy-en-Josas, France) are fully acknowledged for the animal experiments.

References

Ball LA (1992) Cellular expression of a functional novirhabdovirus RNA replicon from vaccinia virus vectors. J Virol 66:2335–2345

Basurco B, Benmansour A (1995) Distant strains of the fish rhabdovirus VHSV maintain a sixth functional cistron which codes for a nonstructural protein of unknown function. Virology 212:741–745

Betts AM, Stone DM (2001) Rescue of viral haemorrhagic septicaemia virus minigenomes by helper virus. Virus Res 77:19–23

Biacchesi S, Thoulouze MI, Béarzotti M, Yu YX, Brémont M (2000a) Recovery of NV knockout Infectious hematopoietic necrosis virus expressing foreign genes. J Virol 74:11247–11253
Biacchesi S, Yu YX, Bearzotti M, Tafalla C, Fernandez-Alonso M (2000b) Rescue of synthetic salmonid rhabdovirus minigenomes. J Gen Virol 81:1941–1945
Biacchesi SM, Bearzotti Bouguyon E, Brémont M (2002) Heterologous exchanges of the glycoprotein and the matrix protein in a Novirhabdovirus. J Virol 76:2881–2889
Collins PL, Mink MA, Stec DS (1991) Rescue of synthetic analogs of respiratory syncytial virus genomic RNA and effect of truncations and mutations on the expression of a foreign reporter gene. Proc Natl Acad Sci U S A 88:9663–9667
Conzelmann KK (1998) Nonsegmented negative-strand RNA viruses: genetics and manipulation of viral genomes. Annu Rev Genet 32:123–162
De BP, Banerjee AK (1993) Rescue of synthetic analogs of genome RNA of human parainfluenza virus type. Virology 196:344–348
Dimock K, Collins PL (1993) Rescue of synthetic analogs of genomic RNA and replicative-intermediate RNA of human parainfluenza virus type 3. J Virol 67:2772–2778
Fearns R, Peeples ME, Collins PL (1997) Increased expression of the N protein of respiratory syncytial virus stimulates minigenome replication but does not alter the balance between the synthesis of mRNA and antigenome. Virology 236:188–201
Fijan N, Sulimanovic M, Béarzotti M, Muzinic D, Zwillenberg LO, Chimonczyk S, Vautherot JF, de Kinkelin P (1983) Some properties of the Epithelioma papulosum cyprini EPC cell line from carp Cyprini carpio. Ann Pasteur/Virologie 134E:207–220
Fuerst TR, Niles EG, Studier FW, Moss B (1986)Eukaryotic transient-expression system based on recombinant vaccinia virus that synthesizes bacteriophage T7 RNA polymerase. Proc Natl Acad Sci U S A 83:8122–8126
Gorman CM, Moffat LF, Howard BH (1982) Recombinant genomes which express chloramphenicol acetyltransferase in mammalian cells. Mol Cell Biol 2:1044–1051
Hoffmann BH, Schutze Mettenleiter TC (2003) Recognition of cis-acting elements of infectious haematopoietic necrosis virus and viral hemorrhagic septicemia virus by homologous and heterologous helper proteins. Virus Res 93: 79–89
Johnson MC, Simon BE, Kim CH, Leong JA (2000) Production of recombinant snakehead rhabdovirus: the NV protein is not required for viral replication. J Virol 74:2343–2350
Kahn JS, Schnell MJ, Buonocore L, Rose JK (1999) Recombinant vesicular stomatitis virus expressing respiratory syncytial virus (RSV) glycoproteins: RSV fusion protein can mediate infection and cell fusion. Virology 254: 81–91
Kretzschmar E, Buonocore L, Schnell MJ, Rose JK (1997) High efficiency incorporation of functional influenza virus glycoproteins into recombinant vesicular stomatitis viruses. J Virol 71:5982–5989
Kurath G, Leong JC (1985) Characterization of infectious hematopoietic necrosis virus mRNA reveals a nonvirion rhabdovirus protein. J Virol 53:462–468
Lawson ND, Stillman EA, Whitt MA, Rose JK (1995) Recombinant vesicular stomatitis viruses from DNA. Proc Natl Acad Sci U S A 92:4477–4481
Lecocq-Xhonneux FM, Thiry Dheur I, Rossius M, Vanderheijden N, Martial J, de Kinkelin P (1994) A recombinant viral haemorrhagic septicaemia virus glycoprotein expressed in insect cells induces protective immunity in rainbow trout. J Gen Virol 75:1579–1587
Lorenzen N, Olesen NJ (1997) Immunization with viral antigens: viral haemorrhagic septicaemia. Dev Biol Stand 90:201–209

Lorenzen N, Olesen NJ, Jorgensen PE (1990) Neutralization of Egtved virus pathogenicity to cell cultures and fish by monoclonal antibodies to the viral G protein. J Gen Virol 71:561–567

Lorenzen N, Olesen NJ, Jorgensen PE, Etzerodt M, Holtet TL, Thogersen HC (1993) Molecular cloning and expression in *Escherichia coli* of the glycoprotein gene of VHS virus, and immunization of rainbow trout with the recombinant protein. J Gen Virol 74:623–630

Lorenzen NE, Lorenzen Einer-Jensen K, LaPatra SE (2002) Immunity induced shortly after DNA vaccination of rainbow trout against rhabdoviruses protects against heterologous virus but not against bacterial pathogens. Dev Comp Immunol 26:173–179

Mebatsion T, Conzelmann KK (1996) Specific infection of $CD4^+$ target cells by recombinant rabies virus pseudotypes carrying the HIV-1 envelope spike protein. Proc Natl Acad Sci U S A 93:11366–11370

Mebatsion T, Finke S, Weiland F, Conzelmann KK (1997) A CXCR4/CD4 pseudotype rhabdovirus that selectively infects HIV-1 envelope protein-expressing cells. Cell 90:841–847

Morzunov SP, Winton JR et al (1995) The complete genome structure and phylogenetic relationship of infectious hematopoietic necrosis virus. Virus Res 38:175–192

Noonan BP, Enzmann J, Trust TJ (1995) Recombinant infectious hematopoietic necrosis virus and viral hemorrhagic septicemia virus glycoprotein epitopes expressed in Aeromonas salmonicida induce protective immunity in rainbow trout (Oncorhynchus mykiss) Appl Environ Microbiol 61:3586–3591

Park KH, Huang T, Correia FF, Krystal M (1991) Rescue of a foreign gene by Sendai virus. Proc Natl Acad Sci U S A 88:5537–5541

Pattnaik AK, Ball LA, LeGrone AW, Wertz GW (1992) Infectious defective interfering particles of VSV from transcripts of a cDNA clone. Cell 69:1011–1120

Pekosz A, He B, Lamb RA (1999) Reverse genetics of negative-strand RNA viruses: closing the circle. Proc Natl Acad Sci U S A 96:8804–8806

Perrotta AT, Been MD (1991) A pseudoknot-like structure required for efficient self-cleavage of hepatitis delta virus RNA. Nature. 350:434–436

Racaniello VR, Baltimore D (1981) Cloned poliovirus complementary DNA is infectious in mammalian cells. Science 214:916–919

Randhawa JS, Marriott AC, Pringle CR, Easton AJ (1997) Rescue of synthetic minireplicons establishes the absence of the NS1 and NS2 genes from avian pneumovirus. J Virol 71:9849–9854

Schnell MJ, Mebatsion T, Conzelmann KK (1994) Infectious rabies viruses from cloned cDNA. EMBO J 13:4195–4203

Schnell MJ, Buonocore L, Kretzschmar E, Johnson E, Rose JK (1996) Foreign glycoproteins expressed from recombinant vesicular stomatitis viruses are incorporated efficiently into virus particles. Proc Natl Acad Sci U S A 91:11359–11365

Schutze HP, Enzmann J, Kuchling R, Mundt E, Niemann H, Mettenleiter TC (1995) Complete genomic sequence of the fish rhabdovirus infectious haematopoietic necrosis virus. J Gen Virol 76:2519–2527

Schutze H, Mundt E, Mettenleiter TC (1999) Complete genomic sequence of viral hemorrhagic septicemia virus, a fish rhabdovirus. Virus Genes 19:59–65

Sidhu MS, Chan J, Kaelin K, Spielhofer P, Radecke F, Schneider H, Masurekar M, Dowling PC, Billeter MA, Udem SA (1995) Rescue of synthetic measles virus minireplicons: measles genomic termini direct efficient expression and propagation of a reporter gene. Virology 208:800–807

Thoulouze MI, Bouguyon E, Carpentier C, Brémont M (2004) Essential role of the NV protein of *Novirhabdovirus* for pathogenicity in rainbow trout. J Virol 78:4098–4107

Whelan SPJ, Ball LA, Barr JN, Wertz GTW (1995) Efficient recovery of infectious vesicular stomatitis virus entirely from cDNA clones. Proc Natl Acad Sci U S A 92:8388–8392

Winton JR (1997) Immunization with viral antigens: infectious haematopoietic necrosis. Dev Biol Stand 90:211–220

Yunus AS, Krishnamurthy S, Pastey MK, Huang Z, Khattar SK, Collins PL, Samal SK (1999) Rescue of a bovine respiratory syncytial virus genomic RNA analog by bovine, human and ovine respiratory syncytial viruses confirms the "functional integrity" and "cross-recognition" of BRSV cis-acting elements by HRSV, ORSV. Arch Virol 144:1977–1990

CTMI (2005) 292:143–163

Plant Rhabdoviruses

M. G. Redinbaugh[1] (✉) · S. A. Hogenhout[2]

[1]Department of Plant Pathology, ARS Corn and Soybean Research,
Wooster OH, 44691, USA
redinbaugh.2@osu.edu

[2]Department of Entomology, Ohio State University-OARDC,
Wooster OH, 44691, USA

Abstract This chapter provides an overview of plant rhabdovirus structure and taxonomy, genome structure, protein function, and insect and plant infection. It is focused on recent research and unique aspects of rhabdovirus biology. Plant rhabdoviruses are transmitted by aphid, leafhopper or planthopper vectors, and the viruses replicate in both their insect and plant hosts. The two plant rhabdovirus genera, *Nucleorhabdovirus* and *Cytorhabdovirus*, can be distinguished on the basis of their intracellular site of morphogenesis in plant cells. All plant rhabdoviruses carry analogs of the five core genes: the nucleocapsid (N), phosphoprotein (P), matrix (M), glycoprotein (G) and large or polymerase (L). However, compared to vesiculoviruses that are composed of the five core genes, all plant rhabdoviruses encode more than these five genes, at least one of which is inserted between the P and M genes in the rhabdoviral genome. Interestingly, while these extra genes are not similar among plant rhabdoviruses, two encode proteins with similarity to the 30K superfamily of plant virus movement proteins. Analysis of nucleorhabdoviral protein sequences revealed nuclear localization signals for the N, P, M and L proteins, consistent with virus replication and morphogenesis of these viruses in the nucleus. Plant and insect factors that limit virus infection and transmission are discussed.

1 Abbreviations

1.1 Virus acronyms

BYSMV	Barley yellow striate mosaic virus
LNYV	Lettuce necrotic yellows virus
MFSV	Maize fine streak virus
MIMV	Maize Iranian mosaic virus
MMV	Maize mosaic virus
NCMV	Northern cereal mosaic virus
PYDV	Potato yellow dwarf virus
RTYV	Rice transitory yellows virus
SSMV	Sorghum stunt mosaic virus
SYNV	Sonchus yellow net virus
SYVV	Sowthistle yellow vein virus
VSV	Vesiculostomatitis virus
WASMV	Wheat American striate mosaic virus

1.2 Other Abbreviations

ER	Endoplasmic reticulum
G	Glycoprotein
GFP	Green fluorescent protein
L	Large polymerase protein
LRI	Leaf rub inoculation
M	Matrix
ORF	Open reading frames
NLS	Nuclear localization signal
N	Nucleocapsid
P	Phosphoprotein
SEL	Size exclusion limit
TEM	Transmission electron microscopy
VIGS	Virus induced gene silencing
VPI	Vascular puncture inoculation

2 Introduction

In this chapter, we will present an overview of the current information on plant rhabdovirus structure and taxonomy, genome structure, protein function, and insect and plant infection. Rather than being comprehensive, we will focus on research that is either recent or that emphasizes unique aspects of rhabdovirus biology. Many rhabdovirus species have two natural hosts: either insects and plants, or insects and vertebrates (Hogenhout et al. 2003). Plant-infecting rhabdoviruses include a number of economically important pathogens and are transmitted to their plant hosts by insect vectors including aphids, planthoppers and leafhoppers (Jackson 1987). The family *Rhabdoviridae* also contains several human and animal pathogens of which the economically important livestock-infecting vesiculoviruses and ephemeroviruses are also insect transmitted. In fact, the widespread ability of rhabdoviruses to infect arthropods has led to speculation that the family evolved from an ancestral insect virus (Nault 1997) and that the rhabdovirus host range is largely determined by the insect host (Hogenhout et al. 2003).

Most plant viruses are composed simply of nucleic acid encapsidated by a protein coat. However, the composition of rhabdovirus particles is more complex. Like other members of the *Rhabdoviridae*, plant rhabdoviruses are bacilliform and are composed of RNA encapsidated by a protein coat surrounded by a lipid bilayer derived from the plant or insect host by budding from the cell's nuclear, endoplasmic reticulum (ER) or cell surface membranes. The rhabdovirus genomic RNA has negative polarity. Thus, the plant rhabdoviruses must carry proteins in the mature virion that enable synthesis of viral mRNAs after introduction into host cells. At 13–14 kb and encoding six to nine proteins, the plant rhabdovirus genome is a large relative to those of most other plant viruses, which range from about 3 to 20 kb and encode from 1 to 12 proteins (Hull 2002).

The ICTV database (Brunt et al. 1996) lists eight definitive cytorhabdoviruses and 13 nucleorhabdoviruses (Table 1). Plant rhabdovirus species are generally distinguished by serological relationships combined with the virus' plant host range and vector species. These viruses are distributed worldwide. The spread of rhabdovirus diseases in crops is restricted by several factors. Rhabdoviruses may have limited plant host range, and some crop germplasm in otherwise susceptible species can be resistant to systemic rhabdovirus infection. Further, plant rhabdoviruses are usually not transmitted vertically to seed or insect progeny, and subsequently are dependent on acquisition and inoculation by their arthropod host. The efficiency of rhabdovirus transmission by insects can be as low as 5% (Ammar 1994; Sylvester and

Table 1 Plant-infecting rhabdoviruses

Virus species name	Acronym	LRI[a]	Vector family	Vectors
Barley yellow striate mosaic cytorhabdovirus	BYSMV	N	*Delphacidae*	*Laodelphax striatellus*
Broad bean yellow vein cytorhabdovirus	BBYVV	N	–	–
Broccoli necrotic yellows cytorhabdovirus	BNYV	Y	*Aphididae*	*Brevicoryne brassicae*
Cereal northern mosaic cytorhabdovirus	NCMV	N	*Delphacidae*	–
Festuca leaf streak cytorhabdovirus	FLSV	N	–	–
Lettuce necrotic yellows cytorhabdovirus	LNYV	Y	*Aphididae*	*Hyperomyzus lactucae*
Sonchus cytorhabdovirus	SonV	Y	–	–
Strawberry crinkle cytorhabdovirus	SCV	Y	*Aphididae*	*Chaetosiphon* sp.
Cynodon chlorotic streak nucleorhabdovirus	CCSV	N	*Delphacidae*	*Toya propinqua*
Coriander feathery red vein nucleorhabdovirus	CFRVV	Y	*Aphididae*	*Hydaphis foeniculi, Myzus persicae*
Carrot latent nucleorhabdovirus	CLV	N	*Aphididae*	*Brachycaudus heraclei*
Datura yellow vein nucleorhabdovirus	DYVV	N	*Not Aphididae*	–
Eggplant mottled dwarf nucleorhabdovirus	EMDV	Y	–	–
Maize fine streak nucleorhabdovirus	MFSV	N	*Cicadellidae*	*Graminella nigrifrons*
Maize mosaic nucleorhabdovirus	MMV	N	*Delphacidae*	*Peregrinus maidis*
Pittosporum vein yellowing nucleorhabdovirus	PVYV	Y	*Not Aphididae*	–
Potato yellow dwarf nucleorhabdovirus	PYDV	Y	*Cicadellidae*	*Agallia constricta, A. quadripunctata, Aceratagallia sanguinolenta*
Rice transitory yellows virus	RTYV	N	*Cicadellidae*	*Nephotettix* sp.
Sonchus yellow net nucleorhabdovirus	SYNV	Y	*Aphididae*	*Aphis coreopsidis*
Sorghum stunt mosaic virus	SSMV	N	*Cicadellidae*	*Graminella sonora*
Sowthistle yellow vein nucleorhabdovirus	SYVV	N	*Aphididae*	*Hyperomyzus lactucae*
Tomato vein clearing nucleorhabdovirus	TVCV	Y	–	–
Wheat Am. striate mosaic nucleorhabdovirus	WASMV	N	*Cicadellidae*	*Elymana virescens, Endria inimica*

[a] The ability of the virus to be transmitted (Y) or not (N) by leaf rub inoculation (LRI) under laboratory conditions. SCV and PYDV cannot be transmitted to all hosts by LRI.

Richarson 1992), and rhabdovirus plant hosts appear to be mostly annuals, whereas perennial plant hosts provide a long-term and continuous reservoir of virus inoculum. Thus, most rhabdovirus species have limited geographic ranges, largely because of the limited distribution of insect vectors and plant hosts. Nonetheless, losses from rhabdovirus diseases in sweet corn can reach 100% (Brewbaker 1981).

The distinctive morphology of rhabdoviruses allows for easy identification of the virions by transmission electron microscopy (TEM) of negatively stained plant extracts. However, further characterization of plant rhabdoviruses is more difficult, because these viruses are not easy to purify from infected plants. In addition, most plant rhabdoviruses are recalcitrant to mechanical transmission and appear to be obligately transmitted by insect vectors, particularly those transmitted by leafhoppers (*Cicadellidae*) and planthoppers (*Delphacidae*) (Hogenhout et al. 2003). Hence, there are a number of plant-infecting rhabdoviruses for which only preliminary descriptions are available (Jackson et al. 1999). Of the 60 or so potential plant-infecting rhabdoviruses, a few have been characterized at the biochemical and molecular levels. Viruses for which sequence information is available include the nucleorhabdoviruses *Rice transitory yellows nucleorhabdovirus* (RTYV), *Sonchus yellow net nucleorhabdovirus* (SYNV), *Maize mosaic virus* (MMV) and maize fine streak virus (MFSV), and the cytorhabdoviruses *Northern cereal mosaic cytorhabdovirus* (NCMV) and *Lettuce necrotic yellow cytorhabdovirus* (LNYV).

3
Morphology and Taxonomy

The basic structure and morphology of plant rhabdoviruses in electron micrographs is similar to that of vertebrate rhabdoviruses. The virions are bacilliform with a spiked surface and a striated capsid core. Virion sizes vary depending on the virus species and fixation methods with lengths of 130–350 nm and widths of 45–100 nm (Jackson et al. 1999). Rhabdovirus virions are composed of a negative-strand RNA genome encapsidated by the nucleocapsid (N) protein surrounded by a phospholipid membrane. The lipid-embedded glycoprotein (G) and matrix (M) proteins are also structural components of the virion. The N, G and M proteins are generally detectable after SDS-PAGE of purified virions. The phosphoprotein (P) and large polymerase (L) proteins, together with the N protein are required for mRNA synthesis and genome replication. These two proteins are present in the virion, but are usually detectable only by serological analyses (Jackson 1987; Wagner et al. 1996).

The plant-infecting rhabdoviruses are separated into the two genera, *Nucleorhabdovirus* and *Cytorhabdovirus*, depending on the intracellular site of virus maturation in plant cells (Walker et al. 2000). Nucleorhabdoviruses accumulate within the nucleus and perinuclear space of the plant and some insect cells (Ammar and Nault 1985; McDaniel et al. 1985). Cytorhabdoviruses appear to replicate solely within the cytoplasm, undergoing morphogenesis in the ER in masses of thread-like structures that are often referred to as viroplasms. However, the distinction between cytorhabdoviruses and nucleorhabdoviruses is not always clear, as budding from nuclear membranes has been observed for some cytorhabdoviruses such as LNYV (Dietzgen et al. 1994). The two plant rhabdovirus genera cannot be distinguished on the basis of plant host or insect vector range (Table 1). For example, both genera contain species that infect maize and sowthistle (*Sonchus arvensis*). In addition, examples of aphid-, leafhopper- and planthopper-transmitted viruses are found in both virus genera.

In insects, the site of rhabdovirus morphogenesis appears to be cell-type-specific. The nucleorhabdovirus *Maize mosaic virus* (MMV) buds primarily from inner nuclear membranes of most tissues in both its planthopper vector (*Peregrinus maidis)* and in maize cells (Ammar and Nault 1985; McDaniel et al. 1985). However, in salivary gland secretory cells, MMV particles commonly bud from the plasma membranes, and accumulate in the intercellular spaces. These spaces ultimately connect to the salivary ducts, where the virus can move to a new plant host via the saliva.

4
Plant Rhabdovirus Genome Organization

At 13–14 kb, plant rhabdoviruses genomes are somewhat larger than most rhabdoviral genomes (Walker et al. 2000). Each plant virus encodes homologs of the five structural rhabdoviral genes found in the prototypical *Vesicular stomatitis virus* (VSV) genome (Fig. 1). The five structural genes occur in the same order as in VSV, and are thought to serve analogous roles in plant rhabdoviruses (Jackson et al. 1999). In addition to these five genes, nucleorhabdovirus and cytorhabdovirus genomes carry one to four additional open reading frames (ORF).

The intergenic regions of rhabdoviruses are highly conserved both among genes of a virus species and between members of the virus family (Table 2). Similarly to VSV intragenic sequences, plant rhabdovirus intergenic regions are comprised of an AT-rich region (region I), thought to be the intracellular polyadenylation signal or stutter sequence, followed by a short, variable, non-

SYNV	**(N)**	N	P	*sc4*				M	G		L
RTYV	**(N)**	N	P	3				M	G	6	L
NCMV	**(C)**	N	P	3	4	5	6	M	G		L
LNYV	**(C)**	N	P	4b				M	G		L
VSV		N	P					M	G		L

Fig. 1 Organization of plant rhabdovirus genomes. The genes and detected open reading frames (ORF) are indicated for the sequenced nucleorhabdoviruses (*N*), cytorhabdoviruses (*C*) and vesiculostomatitis virus (*VSV*). The five core genes found in VSV encode the nucleocapsid (*N*), phosphoprotein (*P*), matrix protein (*M*), glycoprotein (*G*) and the large or polymerase (*L*) protein. The location of the *sc4* and 4b genes and other ORF (shown as *numbers*) found in plant rhabdoviruses are indicated. The plant rhabdovirus sequences used are as indicated in Table 2

transcribed region (region II) and a highly conserved transcription start site (region III). The MFSV intergenic sequence shows a high degree of similarity to those of SYNV and RTYV (Tsai et al., in press). Although there is information available for only a few plant rhabdovirus species, it appears that the region I sequences are more conserved among the nucleorhabdoviruses and that region II is longer in the two cytorhabdoviruses. These patterns suggest conservation of intragenic regions within a genus.

As expected, the L genes are the most conserved of the five core rhabdovirus genes. There is 40%–60% sequence identity among the plant rhabdoviruses, and 37%–39% identity between plant rhabdoviruses and VSV, as determined by ClustalX alignment (Thompson et al. 1997). The other core rhabdovirus genes are somewhat less conserved, with 30%–48% identity among plant rhabdoviruses, and 30%–40% identity between the plant rhabdoviruses and VSV. Phylogenetic analyses of aligned sequences using a neighbor-joining dis-

Table 2 Conservation of plant rhabdovirus intergenic region sequences

Virus[a]	I[b]	II	III
SYNV	AUAUAAGAAAAA	CC	AAC
RTYV	AUAUAAUAAAAA	CCC	AAC
NCMV	AAUUAAGAAAAA	CUGAGAUC	AAC
LNYV	GAUUUAAGAAAA	CNACTGA	GAA
VSV	NUAUGAAAAAAA	CU	AAC

[a]SYNV, *Sonchus yellow net virus* (Genbank L32603); RTYV, *Rice transitory yellows virus* (AB011257); NCMV, *Northern cereal mosaic virus* (AB030227); LNYV, *Lettuce necrotic yellows virus*] (Wetzel et al. 1994); VSV, *Vesiculostomatitis virus* (NC001560).
[b]The consensus sequences for the putative polyadenylation signal (I), the untranscribed intergenic sequence (II) and the known (SYNV, LNYV) or putative transcription start site (III) indicated are from the + strand. Nucleotides conserved for all genes in a viral genome are underlined.

tance algorithm (Swofford 2002) indicate that the three nucleorhabdoviruses (SNYV, MFSV and RTYV) form a group separate from the cytorhabdoviruses (NCMV and LYNV) for the L and N genes (Redinbaugh et al., unpublished results). These data suggest that the genus distinction between nucleorhabdoviruses and cytorhabdoviruses is reflected in the phylogeny of the species. However, similar phylogenetic distinctions between nucleorhabdoviral and cytorhabdoviral P, M and G genes were not found.

Relative to VSV, each of the plant rhabdoviruses carries at least one additional ORF. Several factors indicate that these are expressed genes. Each of the ORFs is downstream of a conserved intergenic region, and transcripts corresponding to these ORFs can be detected by Northern blot analysis of infected plants or insects (Tanno et al. 2000; Scholthof et al. 1994; Huang et al. 2003; Wetzel et al. 1994; Tsai et al., in press). In addition, the ORF-encoded proteins corresponding to the SYNV *sc4* (Scholthof et al. 1994) and the RTYV ORF 6 (Huang et al. 2003) have been detected in virions, plants and insects.

In the plant rhabdoviruses, at least one of the additional ORFs is located between the P and M genes (Fig. 1). The fact that each virus has at least one gene inserted between P and M in the genome suggests a conserved role for these additional genes. In spite of the conserved configuration, no significant sequence similarity can be detected among these genes. The nucleorhabdovirus SYNV and the cytorhabdovirus LYNV each encode six genes, with the *sc4* (Scholthof et al. 1994) and 4b (Wetzel et al. 1994) genes being inserted into the genome between the P and M genes. The nucleorhabdoviruses RTYV and MFSV both encode seven genes. For MFSV, ORF3 and ORF4 are both located between the P and M genes in the viral genome (Tsai et al., in press), while the RTYV ORF6 is located between the G and L genes in a similar location to the NV gene of novirhabdoviruses (Walker et al. 2000). Finally, the cytorhabdovirus NCMV encodes nine ORFs, with ORFs 3–6 being inserted between the putative P and M genes (Tanno et al. 2000). These four ORFs are quite small, encoding proteins of 12.8–18.6 kDa. Because each ORF is preceded by a conserved intergenic region, and transcripts hybridizing to each ORF accumulated in insects, they appear to be genes (Tanno et al. 2000).

4.1 Plant Rhabdovirus Proteins

Direct experimental evidence for plant rhabdovirus protein function and localization is limited to studies of SYNV (Jackson et al. 1987; Scholthof et al. 1994; Choi et al. 1992; Martins et al. 1998; Goodin et al. 2001, 2002) and MFSV (Tsai et al., in press) For the other plant rhabdoviruses, most studies have involved analysis of the genome-encoded open reading frames and comparison of these proteins with those of other more studied viruses.

Rhabdovirus L proteins show the highest conservation with 20%–25% amino acid sequence identity among the plant rhabdoviruses. The L proteins are all large (220–240 kDa) proteins that carry conserved motifs characteristic of RNA-dependent RNA polymerases. The SYNV L protein has been associated with an RNA polymerase activity in the nuclei of infected plants (Choi 1992; Wagner et al. 1996; Wagner et al. 1997). This L protein forms a complex with the viral N and P proteins that is important for virus replication and protein synthesis (Wagner et al. 1997). Most plant rhabdovirus L proteins are positively charged at neutral pH. The predicted L protein sequences for the three nucleorhabdoviruses each have consensus nuclear localization signals (NLS) and are predicted to be localized in the nucleus based on PSORTII analysis (Table 3). Nuclear localization signals are also found in BEFV and the novirhabdovirus L proteins.

Although the predicted plant rhabdovirus N proteins show only about 20% sequence identity, all are basic (pI 8–10) relative to VSV (pI 6.3). Immunocytochemical analyses and the distribution of GFP:N fusion proteins both indicate that the SYNV N protein is targeted to the nucleus (Martins et al. 1998: Goodin et al. 2001). This protein carries a bipartite NLS between amino acids 441 and 461 in the carboxyterminus that is required for nuclear localization (Goodin et al. 2001). PSORTII analyses of the N proteins of the nucleorhabdoviruses RTYV and MFSV also indicate the presence of strong NLS and predict a nuclear localization. Further, the MFSV N protein is also directed to the nucleus (Tsai et al., in press). Neither cytorhabdovirus N protein has a predicted NLS, and none were identified among other rhabdoviruses.

The plant rhabdovirus P proteins have little homology to one another or to other rhabdovirus P proteins. The SYNV P (or M2) protein is found in the virus nucleocapsid core and is associated with the nuclear replicase complex (Wagner et al. 1997). Although immunohistochemical analyses indicate a nuclear localization for the protein (Martins et al. 1998), a GFP:P fusion protein expressed in tobacco cells is distributed between the cytoplasm and nucleus (Goodin et al. 2001). The SYNV P protein does not contain a canonical arginine-lysine-rich NLS, but the protein does have a karyophilic region between amino acids 40 and 124, suggesting that P utilizes an alternative nuclear import pathway (Goodin et al. 2001). Another facet of the complex distribution of P in plant cells is the fact that a GFP:P fusion protein with a carboxyterminal deletion from amino acids 247 and 346 had more pronounced nuclear localization than the wild-type P protein fused to GFP (Goodin et al. 2001). Taken together, the results suggest that P has the ability to shuttle between the nucleus and cytoplasm. Interestingly, the RTYV P protein, but

Table 3 Plant rhabdovirus protein nuclear localization signals predicted with PSORTII[a]

Pro.	SYNV[b]	RTYV	MFSV	NCMV	LNYV
N	446 RKRR[c] 457 KPKK 0.6 nuc.	404 KK LGPPRANAHS RRKEP 0.96 nuc.	435 KR SSDGTGNVSK KKSRK 0.82 nuc.	None	None
P	none	264 RK DSHHYRTVVS RIEKK 0.76 nuc.	None	None	None
M	229 RKRK 266 RKHR 0.6 nuc.	None	206 KRKR 194 KK EDKAEKATTE KRKRQ 0.91 nuc.	None	None
G	590 KKKR 616 RKKK	None	None	None	None
L	1647 KKRP 2054 KPRR 0.80 nucl.	1240 HRRK	None	351 RK FKEIFYMEYF KKNRK 0.7 nuc.	–

[a]Nuclear localization signals were identified using PROSITEII (Horton and Nakai 1999) (http://psort.nibb.ac.jp/).
[b]Virus acronyms are as indicated in Table 2.
[c]The starting amino acid and sequence of the predicted NLS (top line) and the certainty score from PSORTII (bottom line). Only nuclear localization certainties more than 0.5 are listed.

not the MFSV P protein, carries a strong bipartite NLS (Table 3). The MFSV P protein spreads throughout the cell when expressed singly in tobacco (Tsai et al., in press) is also directed to the nucleus (Tsai et al., in press). No P proteins from the other rhabdovirus genera (including cytorhabdoviruses) carry NLS.

Several specific interactions among SYNV proteins are likely to have roles in virus replication and morphogenesis. The N, P and L proteins are thought to interact to form the replicase complex in the plant cell nucleus (Wagner et al. 1997). All three proteins have karyophilic sequences and are localized in plant nuclei. Interactions between N and P occur in yeast two hybrid systems and in vivo (Goodin et al. 2001). Interestingly, co-expression of N and P drastically affected their localization patterns and resulted in a shift of both proteins to a subnuclear region. Similarly, co-expression of MFSV N and P proteins shifted the localization of both proteins to the nucleolus (Tsai et al., in press). Because the karyophillic domain in P is located within the N-binding domain, it is likely that nuclear transport for the two proteins occurs independently and N–P interactions occur after nuclear import (Goodin et al. 2001).

Similarly with the P protein, the plant rhabdovirus M proteins do not share significant sequence homology. As in other rhabdoviruses, these basic proteins are thought to play a role in nucleocapsid coiling and interaction with the G protein (Jackson et al. 1999). Fluorescent protein fusions of the SYNV and MFSV M proteins localized to the nucleus (Goodin et al. 2001, Tsai et al., in press), and both proteins contain NLS and are predicted by PSORTII to be localized to the nucleus (Table 3). The MFSV M protein, but not that of RTYV, also appears to carry strong NLS. The M proteins from cytorhabdoviruses do not carry NLS, whereas novirhabdovirus (*Snakehead rattle virus* and *Viral hemorrhagic septicemia virus*) M proteins do.

Plant rhabdovirus G proteins are glycosylated and form the virion spikes (Jackson et al. 1999). In SYNV-infected tobacco cells, inhibiting glycosylation with tunicamycin prevented virion morphogenesis and resulted in accumulation of nucleocapsid cores at the periphery of the nucleus (van Beek et al. 1985). The plant rhabdovirus G proteins do not share significant homology with each other, or with other rhabdovirus G proteins. However, all five plant G proteins are predicted to have one (SYNV, MFSV and NCMV) or two (RTYV and LYNV) transmembrane domains, consistent with their role as viral membrane glycoproteins. The SYNV G protein has a NLS, while the other rhabdoviruses do not (Table 3). In addition, immunofluorescence localization with SYNV virion antibodies suggests that most of the virion G protein is localized in the periphery of the nucleus (Martins et al. 1998).

Specific roles for the additional genes in plant rhabdoviruses remain to be investigated. A role in vector transmission is less likely, because VSV is capable of replicating in animals with just the five core genes. Further, VSV can be transmitted by sandflies and blackflies, although it is not yet known whether the virus spreads throughout these insects similarly to the spread of plant rhabdoviruses in their aphid, leafhopper and planthopper vectors. A more likely role for the additional genes is in facilitating cell-to-cell and/or systemic movement in plants. Movement proteins involved in cell-to-cell virus transport have been described for many plant viruses, including the NS_M protein of the thrips-transmitted plant pathogenic tospoviruses that belong to the *Bunyaviridae* (Storms et al. 2002). Interestingly, the human and animal infecting members of the *Bunyaviridae* do not have the NS_M protein.

In contrast to animal systems in which viruses enter cells by receptor-mediated endocytosis, movement of plant viruses between plant cells occurs through the symplast, the continuous protoplasm of plants that interconnects cells via plasmodesmata. Many plant viruses encode proteins that function to increase the size exclusion limit (SEL) of plasmodesmata, allowing relatively large molecules to move between cells (Oparka and Roberts 2000). Since rhabdoviruses are more than ten times larger than the normal 3-nm diameter of plasmodesmata and the viral genome is not by itself infectious; it is highly likely that a plant rhabdovirus protein(s) must function to increase the plasmodesmata SEL so that unenveloped nucleocapsid cores may move from cell to cell. The *sc4* protein is proposed to have a role in cell-to-cell movement. The protein is expressed in plants and binds loosely to the virion (Scholthof et al. 1994). When expressed as a GFP: *sc4* fusion protein, it is partially localized at the plant cell periphery (Goodin et al. 2002). In addition, the protein is related to the 30K superfamily of plant virus movement proteins (Melcher 2000). Interestingly, the predicted LNYV 4b protein has similarity to the capillovirus and trichovirus movement proteins, which are in a different group of movement proteins in the 30K superfamily from *sc4* (U. Melcher, personal communication).

Another possible function of the additional genes in plant rhabdoviruses is to manipulate or suppress plant defense responses such as virus-induced gene silencing (VIGS) (Melcher 2000; Voinnet 2001). VIGS involves sequence-specific degradation of viral RNA transcripts or genomes, and spreads systemically throughout the plant in response to localized virus infection (Vionnet 2001). Many plant viruses have evolved counter-defensive proteins against VIGS known as suppressors of gene silencing. These include the helper component-protease of potyviruses and the *Cucumber mosaic virus 2b* protein (Vionnet et al. 1999). Many of the identified viral suppressors of gene silencing were previously described as having phenotypes involved with systemic or long-distance virus movement in the plant. Although most of the viral suppressors of gene silencing have been identified from positive-strand

RNA viruses, negative-strand RNA virus genomes also encode suppressors. The tospovirus and tenuivirus NS_S proteins serve this function (Bucher et al. 2003). Although no suppressor of gene silencing has been identified in plant rhabdoviruses, RTYV is capable of inducing VIGS in rice as virus resistance is seen in transgenic rice expressing RTYV N gene sequences (Fang et al. 1996).

Recently, it was discovered that that the *Drosophila*-infecting *Flock house virus* both induces gene silencing and encodes a suppressor of silencing that is active in the insect host (Li et al. 2002). This raises the possibility that insect-transmitted plant rhabdoviruses manipulate the gene silencing system of their insect hosts in order to be transmitted. It is unclear whether and how gene silencing plays a role in organisms that have an adaptive immune response; however, it seems likely that most, if not all, plant-infecting viruses, including rhabdoviruses, have suppressors of gene silencing.

Other than *sc4*, the only additional plant rhabdovirus gene that was shown to be expressed is the RTYV ORF 6. This gene encodes a small (10.5-kDa) phosphorylated, acidic protein with limited sequence similarity to the novirhabdovirus non-virion proteins (Huang et al. 2003). The protein can be detected in the virion and leafhopper vector, but not in infected rice, suggesting a role in insect transmission for this protein.

5 Virus Infection of Plant Hosts

Insects are thought to inoculate plants with rhabdoviruses in nature by injecting the virus into the plant cell through wounds produced by their stylets. In the laboratory, some rhabdoviruses can be transmitted using mechanical inoculation techniques such as leaf rub inoculation (LRI) or vascular puncture inoculation (VPI) that may mimic the inoculations by vectors as they use their stylets to probe leaf epidermal cells prior to feeding (Hull 2002, Hogenhout et al. 2003). For LRI, a virus-containing solution containing an abrasive (e.g., Celite) is rubbed on the leaf surface, making wounds in the epidermal cells and allowing for virus entry. LRI works well for many of the aphid-transmitted rhabdoviruses including SYNV (Table 1). Most of the rhabdoviruses transmitted by planthoppers and leafhoppers are recalcitrant to LRI, indicating the virus must be delivered to specific, nonepidermal cells for infection to begin (Nault and Ammar 1989). This is consistent with the feeding pattern of leafhoppers and planthoppers, as they generally move their stylets between cells until they find the sugar-rich phloem cells. However, leafhopper- and planthopper- transmitted maize rhabdoviruses can be mechanically transmitted using VPI, which uses a jeweler's engraving tool to drive minuten pins through virus inoculum and into the scutellum of germinating maize kernels (Louie 1995). It is thought that this technique produces wounds in those specific cells that are susceptible to infection by the viruses (Hogenhout et al. 2003).

Inoculation efficiency of plant rhabdoviruses is strongly affected by the level of resistance the plant has to the virus. Plants, like vertebrates and arthropods, have an innate immune response that is triggered by microbial infection, and there is evidence that this system is important in the plant response to virus infection (Nurnberger and Scheel 2001). For example, the N gene of tobacco confers resistance to tobacco mosaic virus in a classic gene-for-gene manner, by inducing a hypersensitive response and rapid cell death at the site of infection when plants carrying the N gene are inoculated with virus (Whitham et al. 1996). While plant genes conferring resistance to rhabdoviruses have yet to be isolated, there is genetic evidence for rhabdovirus resistance in maize (Ming et al. 1997; Redinbaugh et al. 2001).

Although rhabdoviruses can be transmitted mechanically, the insect is likely to be more than an injection needle, because insect saliva may play a significant role in the establishment of rhabdovirus infection in vertebrate and plant hosts. Insects are much more efficient transmitters of rhabdoviruses, requiring significantly lower concentrations of virus to establish infections than are needed for mechanical transmission. This may be due, at least in part, to factors such as P-450-monooxygenases, glutathione S transferases and proteases in insect saliva that suppress plant defense responses (Foissac et al. 2002; Feyereisen 1999; Gatehouse 2002). In addition, insects produce and excrete proteases that are insensitive to the protease inhibitors secreted by plants in the defense response (Gatehouse 2002). This modulation of the host defense response might facilitate the initial replication of the virus and thus enhance infectivity.

6
Virus Infection of Insect Hosts

Plant rhabdoviruses are transmitted by cicadellid leafhoppers, delphacid planthoppers and aphids in a persistent propagative manner. For transmission to occur, the insect must acquire virus from a plant host over a period of hours to days, then be retained in the insect for a period of days to weeks prior to transmitting the virus to plants (Ammar and Nault 2002). The insect can transmit the virus for the remainder of its lifetime. The virus replicates within the insect vector, as demonstrated by serial dilution transmission assays, injection experiments and quantitative serology.

To be transmitted by the insect, viruses must cross the cells of the gut, move from the gut to the salivary gland, cross the salivary gland cell layers, move into the saliva and be introduced into the plant or animal host by the insect with the saliva (reviewed in Hogenhout et al. 2003). This path suggests that receptor proteins in the brush border or apical membrane of the insect digestive tract are the first insect cellular molecules with which plant rhabdovirus particles interact (Adam and Hsu 1984; Jackson et al. 1987). It is likely that the virions interact with virus receptors on these membranes and enter cells by receptor-mediated endocytosis in a manner similar that described for VSV and rabies virus (Lewis and Lentz 1998; Superti et al. 1987). From the gut cells, viruses move to other insect organs and tissues including muscle, nervous tissues and into the hemolymph (Ammar and Nault 1985). It is generally thought that salivary gland cells are infected from the insect hemolymph; however, it is also possible that the salivary gland is infected via other tissues such as muscle cells or neurons (Hogenhout et al. 2003, Ammar et al. 2005b). The latter possibility is more in agreement with the spread of lyssaviruses and vesiculoviruses through host nervous tissue prior to infection of the salivary glands (Tyler and McPhee 1987). Nonetheless, virus entry into muscle, neuron, salivary gland and other cells is also likely to occur by receptor-mediated endocytosis. After replication in salivary gland cells, viruses are secreted into the insect saliva an introduced into animal or plant hosts.

The assembly and accumulation sites of maize mosaic virus (MMV) in its planthopper host *P. maidis* were determined by transmission electron microscopy (TEM) (Ammar and Nault 1985). MMV particles were found in most insect organs. In the insect head, virions were observed in the brain, epidermis, fat and connective tissue, retinula cells, muscle and trachea. Virions also were observed in the principal and accessory salivary glands, nerve ganglia, muscle, foregut, midgut, male and female reproductive systems, but not in hindgut cells or Malphigian tubules. As mentioned above, the intracellular accumulation and assembly sites of MMV varied among vector tissues, with MMV budding from inner nuclear membranes of most cell types. In addition to budding from the ER of the secretory cells of the salivary gland, the virus accumulated in large vesicles in the cytoplasm in some nerve cells.

7
Factors Affecting Movement Between Insect and Plant Hosts

There is a difference between an insect being a virus host and a virus vector. To be transmitted to a new plant host by an insect the rhabdovirus must traverse a number of barriers in the insect host (Ammar 1994; Jackson et al. 1999; Hogenhout et al. 2003). These barriers include being acquired from the plant host, entry into and infection of midgut cells, release from midgut cells, movement through the insect hemolymph or other organs, infection of the salivary gland, release of the rhabdovirus into the saliva, inoculation into a new susceptible plant host, and establishment of systemic infection in the new plant host. Blockage at any step may prevent transmission of viruses to plants.

There are at least five serologically distinct rhabdoviruses that infect maize, and each of these viruses has a different planthopper or leafhopper vector. Two strains of *Potato yellow dwarf virus* (PYDV) are differentially transmitted by *Aceratagallia* and *Agallia* species (Adams and Hsu 1984), and two serologically related maize rhabdoviruses, *Sorghum stunt mosaic virus* (SSMV) and MFSV, are differentially transmitted by *Graminella* species (Creamer et al. 1997; Redinbaugh et al. 2001). These results indicate there are viral factors that affect insect host range and vectoring capability that are unrelated to the plant host.

Plant rhabdovirus transmission can be blocked in the insect gut, because injection of virus into the hemolymph can increase transmission efficiency and may allow for transmission of viruses by non-vectors. For example, *Maize Iranian mosaic virus* (MIMV), which is serologically related to some MMV isolates (Ammar et al. 2005a), is naturally transmitted by the planthopper *Ribautodelphax notabilis*, but is transmitted at very low efficiency (0.4%–1.6%) by the MMV vector *P. maidis* (Izadpanah 1989). In contrast, *P. maidis* transmits MIMV at 64% efficiency after acquisition by hemolymph injection (Ammar 1994). Similar results were reported for several aphid-transmitted rhabdoviruses (Sylvester and Richardson 1992).

There is also evidence for non-gut barriers to virus transmission in insects. For example, the aphid *Macrosiphum euphorbiae* remains an inefficient vector of *Sowthistle yellow vein virus* (SYVV) when the virus is acquired by injection into the hemolymph (Behncken 1973), indicating that barrier to virus transmission could not be overcome by bypassing the insect gut. In this case, the virus was detected by TEM in various insect tissues, but was not found in the salivary glands. At the molecular level, organ barriers could result from a failure to enter or replicate in cells, a triggering of host defense responses (immune response and apoptosis), failure to move between cells, or failure to exit cells.

8 Conclusions and Future Perspectives

Analysis of genome sequence information suggests that the differential morphogenesis of cytorhabdoviruses and nucleorhabdoviruses is reflected in the phylogeny of the viral N and L genes. However, the differences in genome structure and sequence among the nucleorhabdoviruses are greater than those identified among species of other rhabdovirus genera, and the situation is similar for cytorhabdoviruses. Further phenotypic and genotypic characterization of plant rhabdoviruses will lead to a better understanding of the distinctions among members of these diverse genera. The sequenced nucleorhabdoviruses have relatively divergent plant hosts and include one aphid-, one leafhopper- and one planthopper-transmitted species. To aid in identifying genomic components important for insect vector and plant host specificity, sequence information from plant rhabdovirus species with overlapping plant host ranges and vector specificity would be useful (e.g., MMV/*P. maidis*, MIMV/*R. notabilis*, MFSV/*G. nigrifrons*, WASMV/*Endria inimica*, SSMV/*G. sonora*, NCMV/*Laodelphax striatellus* and BYSMV/*L. striatellus*).

Development of infectious virus clones would greatly facilitate association of functions with specific rhabdovirus proteins in both plants and insects. Expression of infectious clones will require production of a system expressing the rhabdoviral P and L proteins, as has been used for VSV and rabies virus (Pattniak et al. 1992; Mebatsian et al. 1996). More immediately, systems for transient expression of viral proteins in plants such as those developed by Goodin et al. (2001, 2002) will be useful for localizing viral proteins in plants and for elucidating some virus gene functions, especially identification of viral suppressors of gene silencing.

Insect vectors play a critical role in determining the host range of vertebrate and plant rhabdoviruses, and multiple barriers to rhabdovirus transmission have been identified in insects. Characterization of these barriers at the molecular level will aid in elucidation of the pathway the virus takes through the insect from the gut to the salivary gland before being transmitted to a new plant host. A major question is whether movement of virus through the insect nervous system is required for vector activity. Further, identification of rhabdovirus receptors in the vector midgut and salivary gland may provide information on vector specificity, as would some information on the nature of the insect immune response to rhabdovirus infection. While rhabdoviruses do not generally cause disease in their insect vectors, it will be interesting to determine whether virus infection produces changes in behavior that affect virus transmission to plant hosts.

In agricultural systems, plant resistance to pathogens is usually the most economically feasible and environmentally sustainable method for controlling disease. With the exception of maize, naturally occurring rhabdovirus resistance has not been identified in plant hosts. Therefore, further characterization of the genetic resistance in other crops affected by rhabdovirus diseases is warranted. In addition, characterization of the rhabdovirus resistance response in maize will aid in identification and development of virus-resistant germplasm for a variety of plants.

Acknowledgements We thank Chi-Wei Tsai (Ohio State University), Michael Goodin (University of Kentucky) and Ullrich Melcher (Oklahoma State University) for sharing unpublished results. We thank Tea Meulia, Roy Gingery and Xiaodong Bai for helpful suggestions. Research on the MFSV sequence was supported, in part, by USDA NRI grant 2002–35302–12653.

References

Adam G, Hsu HT (1984) Comparison of structural proteins from 2 potato yellow dwarf viruses. J Gen Virol 65:991–994

Ammar ED (1994) Propagative transmission of plant and animal viruses by insects: factors affecting vector specificity and competence. Adv Dis Vector Res 10:289–331

Ammar ED, Nault LR (1985) Assembly and accumulation sites of maize mosaic virus in its planthopper vector. Intervirology 24:33–41

Ammar ED, Nault LR (2002) Virus transmission by leafhoppers, planthoppers and treehoppers (Auchenorrhyncha, Homoptera). Adv Bot Res 36:141–167

Ammar ED, Gomez-Luengo RG, Gordon DT, Hogenhout SA (2005a) Characterization of Maize Iranian mosaic virus and comparison with Hawaiian and other isolates of Maize mosaic virus (Rhabdoviridae). J. Phytopath in press.

Ammar ED, Meulia T, Özbek E, Hogenhout SA (2005b) Assembly and accumulation sites of Maize mosaic virus (Rhabdoviridae) in plant host and insect vector using transmission electron and confocal laser scanning microscopy. In: Current Issues in Multidisciplinary Microscopy Research and Education, Microscopy book series, Vol. II, Formatex Research Center Badajoz, Spain, in press.

Behncken GM (1973) Evidence of multiplication of sowthistle yellow vein virus in an inefficient aphid vector, Macrospihum euphorbiae. Virology 53:405–412

Brewbaker JL (1981) Resistance to maize mosaic virus. In: Gordon DT, Knoke JK, Scott GE (eds) Virus and viruslike diseases of maize in the United States. So Coop Ser Bull, Wooster, OH. pp 145–151

Brunt A, Crabtree K, Dallwitz M, Gibbs A, Watson L, Zurcher E (1996). Plant viruses online: descriptions and lists from the VIDE database. http://biology.anu.edu.au/Groups/MES/vide/

Bucher E, Sijen T, De Haan P, Goldbach R, Prins M (2003) Negative-strand tospoviruses and tenuiviruses carry a gene for a suppressor of gene silencing at analogous genomic positions. J Virol 77:1329–1336

Choi TJ, Kuwata S, Koonin EV, Heaton LA, Jackson AO (1992) Structure of the L (polymerase) protein gene of sonchus yellow net virus. Virology 189:31–39

Creamer R, He X, Styer WE (1997) Transmission of sorghum stunt mosaic rhabdovirus by the leafhopper vector, *Graminella sonora* (Homoptera: Cicadellidae). Plant Dis 81:63–65

Dietzgen RG (1994) Rhabdoviridae. In: Kohmoto K, Singh RP, Singh US, Zeigler R (1994) Pathogenesis and host-parasite specificity in plant diseases: histopathological, biochemical, genetic and molecular basis 3. Pergmon Press, Oxford

Fang, R-X, Zhu H-T, Wang Q, Mang K-Q, Gao D-M, Qin W-S, Zhang L, Cao S-Y, Tian W-Z, Li L-C (1996) Construction of transgenic rice plant resistant to rice yellow stunt virus, a plant rhabdovirus. In: Rice genetics III. Proceedings of the Third International Rice Genetics Symposium. IRRI, Manila, pp 201–205

Feyereisen R (1999) Insect P450 enzymes. Annu Rev Entomol 44:507–533

Foissac X, Edwards MG, Du JP, Gatehouse AMR, Gatehouse JA (2002) Putative protein digestion in a sap-sucking homopteran plant pest (rice brown plant hopper; *Nilaparvata lugens*: Delphacidae) – identification of trypsin-like and cathepsin B-like proteases. Insect Biochem Mol Biol 32:967–978

Gatehouse JA (2002) Plant resistance towards insect herbivores: a dynamic interaction. New Phytol 156:145–169

Goodin MM, Austin J, Tobias R, Fujita M, Morales C, Jackson AO (2001) Interactions and nuclear import of the N, P proteins of sonchus yellow net virus, a plant nucleorhabdovirus. J Virol 75:9393–9406

Goodin MM, Dietzgen RG, Schichnes D, Ruzin S, Jackson AO (2002) pGD vectors: versatile tools for the expression of green and red fluorescent protein fusions in agroinfiltrated plant leaves. Plant J 31:375–383

Hogenhout SA, Redinbaugh MG, Ammar ED (2003) Plant and animal rhabdovirus host range: a bug's view. Trends Microbiol 11:264–271

Horton P, Nakai K (1996) A probabilistic classification system for predicting the cellular localization sites of proteins. Intell Syst Mol Biol 4:109–115

Huang YW, Zhao H, Luo ZL, Chen XY, Fang RX (2003) Novel structure of the genome of rice yellow stunt virus: identification of the gene 6-encoded virion protein. J Gen Virol 84:2259–2264

Hull R (2001) Matthew's plant virology. Academic Press, New York

Izadpanah K (1989) Purification and serology of the Iranian maize mosaic rhabdovirus. J Phytopathol 126:43–50

Jackson AO (1987) Biology, structure, and replication of plant rhabdoviruses. In: Wagner RR (ed) The rhabdoviruses. Plenum Press, New York, pp 427–508

Jackson AO, Goodin M, Moreno I, Johnson J, Lawrence DM (1999) Plant rhabdoviruses. In: Granoff A, Wagner RG (eds) Encyclopedia of virology, Vol. 3. Academic Press, New York, pp 1531–1541

Jones RW, Jackson AO (1990) Replication of sonchus yellow net virus in infected protoplasts. Virology 179:815–820

Lewis P, Lentz TL (1998) Rabies virus entry into cultured rat hippocampalneurons. J Neurocytol 27:559–573

Li H, Li WX, Ding SW (2002) Induction and suppression of RNA silencing by an animal virus. Science 296:1319–1321

Louie R (1995) Vascular puncture of maize kernels for the mechanical transmission of maize white line mosaic virus and other viruses of maize. Phytopathology 85:139–143

Martins CRF, Johnson JA, Lawrence DM, Choi TJ, Pisi AM, Tobin SL, Lapidus D, Wagner JDO, Ruzin S, McDonald K, Jackson AO (1998) Sonchus yellow net rhabdovirus nuclear viroplasms contain polymerase-associated proteins. J Virol 72:5669–5679

McDaniel LL, Ammar ED, Gordon DT (1985) Assembly, morphology, and accumulation of a Hawaiian isolate of maize mosaic-virus in maize. Phytopathology 75:1167–1172

Mebatsion T, Schnell MJ, Cox JH, Finke S, Conzelmann K-K (1996) Highly stable expression of a foreign gene from rabies virus vectors . Proc Nat. Acad Sci U S A 93:7310–7314

Melcher U (2000) The '30K' superfamily of viral movement proteins. J Gen Virol 81:257–266

Ming R, Brewbaker JL, Pratt RC, Musket TA, McMullen MD (1975) Molecular mapping of a major gene conferring resistance to maize mosaic virus. Theor Appl Genet 95:271–275

Nault LR (1997) Arthropod transmission of plant viruses: a new synthesis. Ann Entomol Soc Amer 90:521–541

Nault LR, Ammar E (1989) Leafhopper and planthopper transmission of plant-viruses. Annu Rev Entomol 34:503–529

Nurnberger T, Scheel D (2001) Signal transmission in the plant immune response. Trends Plant Sci 6:372–379

Oparka KJ, Roberts AG (2001) Plasmodesmata. A not so open-and-shut case. Plant Physiol 125:123–126

Pattnaik AK, Ball LA, Legrone AW, Wertz GW (1992) Infectious defective interfering particles of VSV from transcripts from transcripts of a cDNA clone. Cell 69:1011–1020

Redinbaugh MG, Seifers DL, Abt JJ, Anderson RJ, Styer WE, Ackerman J, Meulia T, Salomon R, Houghton W, Creamer R, Gordon DT, Hogenhout SA (2002) Maize fine streak virus, a new leafhopper-transmitted maize rhabdovirus. Phytopathology 92:1167–1174

Scholthof KBG, Hillman BI, Modrell B, Heaton LA, Jackson AO (1994) Characterization and detection of *sc4* - a 6th gene encoded by sonchus yellow net virus. Virology 204:279–288

Storms MMH, Nagata T, Kormelink R, Goldbach RW, Van Lent JWM (2002) Expression of the movement protein of tomato spotted wilt virus in its insect vector *Frankliniella occidentalis*. Arch Virol 147:825–831

Superti F, Seganti L, Ruggeri FM, Tinari A, Donelli G, Orsi N (1987) Entry pathway of vesicular stomatitis virus into different host cells. J Gen Virol 68:387–399

Swofford DL (2002) PAUP* phylogenetic analysis using parsimony (*and Other Methods). Version 4. Sinauer Associates, Sunderland, MA

Sylvester ES, Richardson J (1992) Aphid-borne rhabdoviruses-relationship with their aphid vectors. In: Harris KF (ed) Advances in disease vector research, Vol. 9. Springer-Verlag, Berlin Heidelberg New York, pp 313–341

Tanno F, Nakatsu A, Toriyama S, Kojima M (2000) Complete nucleotide sequence of northern cereal mosaic virus and its genome organization. Arch Virol 145:1373–1384

Thompson JD, Gibson TJ, Plewniak F, Jeanmougin F, Higgins DG (1997) The ClustalX windows interface: flexible strategies for multiple sequence alignment aided by quality analysis tools. Nuc Acids Res 24:4876–4882

Tsai C-W, Redinbaugh MG, Willie KJ, Reed S, Goodin M, Hogenhout SA Complete genome sequence and in planta subcellular localization of maize fine streak virus proteins. J. Virol., in press.

Tyler KL, McPhee DA (1987) Molecular and genetic aspects of the pathogenesis of viral infections of the central nervous system. Crit Rev Neurobiol 3:221–243
Van Beek NAM, Lohuis D, Dijkstra J, Peters D (1985) Morphogenesis of sonchus yellow net virus in cowpea protoplasts. J Ultrastruct Res 90:294–303
Voinnet O (2001) RNA silencing as a plant immune system against viruses. Trends Genet 17:449–459
Voinnet O, Pinto YM, Baulcombe DC (1999) Supression of gene silencing: a general strategy used by diverse DNA, RNA viruses of plants. Proc Nat Acad Sci U S A 96:14147–14152
Wagner JDO, Choi TJ, Jackson AO (1996) Extraction of nuclei from sonchus yellow net rhabdovirus-infected plants yields a polymerase that synthesizes viral mRNAs and polyadenylated plus-strand leader RNA. J Virol 70:468–477
Wagner JDO, Jackson AO (1997) Characterization of the components and activity of sonchus yellow net rhabdovirus polymerase. J Virol 71:2371–2382
Walker PJ, Benmansour A, Diezgen R, Fang R-X, Jackson AO, Kurath G, Leong JC, Nadin-Davies S, Tesh RB, Tordo N (2000) Family rhabdoviridae. In: Van Regenmortel MHV (ed) Virus taxonomy: 7th report of the International Committee on Taxonomy of Viruses. Academic Press, San Diego, pp 563–583
Wetzel T, Dietzgen RG, Dale JL (1994) Genomic organization of lettuce necrotic yellows rhabdovirus. Virology 200:401–412
Whitham S, McCormick S, Baker B (1996) The N gene of tobacco confers resistance to tobacco mosaic virus in transgenic tomato. Proc Natl Acad Sci U S A 93:8776–8781

CTMI (2005) 292:165–200

Recombinant Rhabdoviruses: Vectors for Vaccine Development and Gene Therapy

S. Finke (✉) · K.-K. Conzelmann

Max von Pettenkofer-Institut & Genzentrum, Ludwig-Maximilians-Universität, Feodor-Lynen-Str. 25, 81377 Munich, Germany
finke@lmb.uni-muenchen.de

Abstract The establishment of methods to recover rhabdoviruses from cDNA, so-called reverse genetics systems, has made it possible to genetically engineer rhabdoviruses and to study all aspects of the virus life cycle by introducing defined mutations into the viral genomes. It has also opened the way to make use of the viruses in biomedical applications such as vaccination, gene therapy, or oncolytic virotherapy. The typical gene expression mode of rhabdoviruses, a high genetic stability, and the propensity to tolerate changes in the virus envelope have made rhabdoviruses attractive, targetable gene expression vectors. This chapter provides an overview on the possibilities to manipulate biological properties of the rhabdoviruses that may be important for further development of vaccine vectors and examples of recombinant rhabdoviruses expressing foreign genes and antigens.

1 Introduction

The *Rhabdoviridae* family, together with the *Paramyxoviridae*, *Filoviridae*, and *Bornaviridae* constitute the *Mononegavirales* order, also known as the nonsegmented negative-strand RNA viruses (NNSV). The common feature of these viruses is the organization of their genetic information in the form of a ribonucleoprotein complex (RNP) and the typical mode of its expression. With respect to genome organization and encoded genes, vesicular stomatitis virus (VSV, genus *Vesiculovirus*) and rabies virus (genus *Lyssavirus*) can be regarded the minimal rhabdoviruses (and minimal *Mononegavirales* as well). Their single-stranded, linear genome RNAs of negative sense (i.e., complementary to mRNA) comprise 11–12 kb and five virus protein-encoding genes in the order 3′-N-P-M-G-L-5′ (Fig. 1A). The genome RNA and the complementary antigenome RNA is always tightly encapsidated by the nucleoprotein (N) to form the helical RNP. Only such RNP, but not free RNA, is suitable as a template for the viral RNA-dependent polymerase, which is composed of a large subunit (L) and a phosphoprotein cofactor (P). Two modes of RNA synthesis are distinguished. Transcription starts at the 3′ end of the genomic RNP and involves sequential production of monocistronic mRNAs from genomic RNPs. Replication initiates at the 3′ ends of both genomic and antigenomic RNPs and produces full-length RNPs. For all rhabdoviruses characterized so far, the presence of an RNP along with N, P, and L protein is sufficient to support all types of viral RNA synthesis. Transcription requires the presence of L/P complexes; replication of RNPs in addition requires a constant supply of N/P complexes, which are used for proper encapsidation of the nascent genomic or antigenomic RNA (Fig. 1A).

The two other proteins encoded on the viral genome, the matrix protein (M) and the glycoprotein (G) are structural components of the viral envelope. The M protein has major functions in the assembly and budding of virions. It is responsible for condensation of RNPs into helical structures known as skeletons from electron micrographs, recruitment of the RNPs to the cell membrane, and budding off the enveloped, typically bullet-shaped rhabdovirus particles (Mebatsion et al. 1999). The M layer beneath the membrane is most likely also responsible for proper recruitment of the transmembrane surface glycoprotein G into the envelope. The trimeric G spikes are essential for binding the virus to the surface of a new target cell. After receptor binding and uptake of the virus by endocytosis, the G spikes mediate pH-dependent fusion of the viral and cellular membrane, resulting in release of the RNP into the cytoplasm. As approximately 50 L/P complexes are associated with the incoming viral RNP, primary transcription can ensue, initiating viral gene expression and replication.

These five virus proteins, N, P, M, G, and L, as found in RV and VSV, represent the minimal and sufficient set of proteins to accomplish a mammalian rhabdovirus life cycle. Each is essential: the lack of either one of them halts either replication (N, P, L) assembly of virus particles (M) or entry into a target cell (G). Additional gene products may be encoded by other rhabdoviruses. It is assumed that most of them are nonessential, as demonstrated for the NV

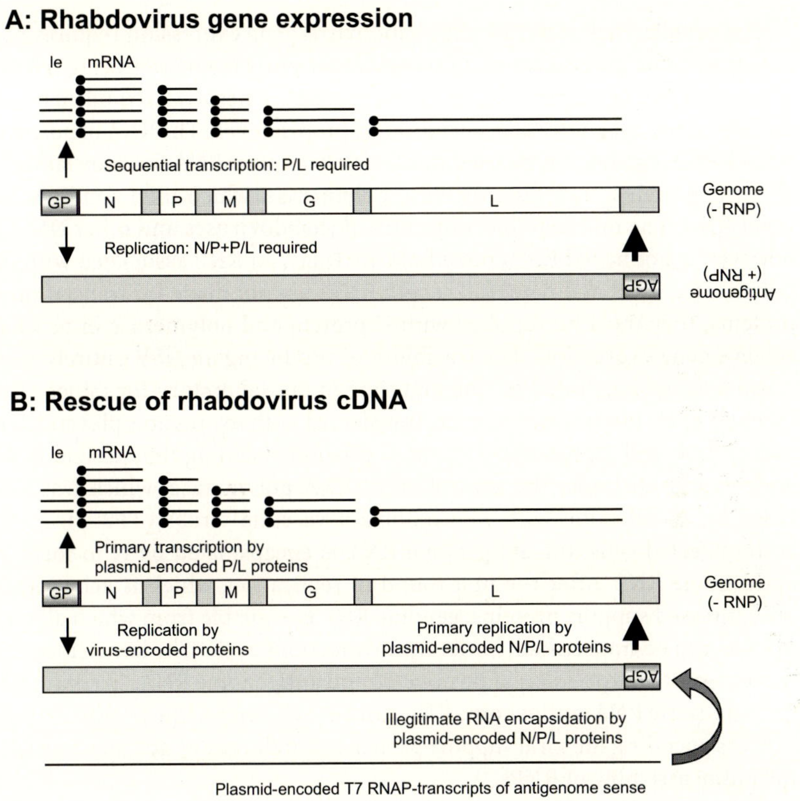

Fig. 1A, B Rhabdovirus gene expression and rescue of rhabdovirus cDNA. **A** The (–)-strand genome RNA is present in a ribonucleo protein complex (–RNP). The –RNP is the template for production of a short leader (le) RNA and sequential transcription of five monocistronic mRNAs. The polymerase enters the genome exclusively at the 3′ promoter (*GP*). Since the polymerase complex P/L eventually dissociates from the template, a transcription gradient is generated. Replicative synthesis of full-length antigenomes, which are concurrently packaged by N/P complexes to +RNPs, is also mediated by the P/L complex. **B** To generate recombinant rhabdoviruses from cDNAs, antigenome sense transcripts are expressed, mostly by bacteriophage T7 RNA polymerase (*T7 RNAP*), in cDNA transfected cells. Coexpression of viral proteins N, P and L essential for virus transcription and replication results in +RNPs and subsequent replication to (–) sense genome RNPs. With expression of all five viral proteins from the genome, an infectious cycle is initiated

protein of novirhabdoviruses. They may have regulatory functions in gene expression or help the virus to better cope with host cell responses. Additional genes of plant rhabdoviruses may be essential with respect to mobility in plants. Such genes are discussed in the other chapters.

The peculiarities of NNSV and rhabdovirus gene expression requires particular systems for generation of recombinant virus from cDNA (Fig. 1B). In contrast to positive-strand RNA viruses, the RNA of negative-strand viruses is not infectious upon introduction into an appropriate cell. The RNA of positive-strand RNA viruses can be used immediately as an mRNA for translation of all virus proteins, including the viral polymerase, which then amplifies the input RNA. The minimal infectious unit of rhabdoviruses and other NNSVs, however, is not naked RNA, but a holo-RNP, i.e., an RNP associated with the polymerase. As rhabdoviral full-length RNAs are not made for translation of proteins, they must be supplied with N protein and polymerase in order to achieve gene expression. The feasibility of producing an NSV entirely from cDNA was demonstrated for the first time in our laboratory for rabies virus (Schnell et al. 1994). Cells were co-transfected with expression plasmids encoding N, P, and L proteins, and with a plasmid encoding the full-length RV antigenome, all under the control of T7 RNA polymerase, which was provided by an infection with a recombinant vaccinia virus (vTF7–3). In the co-transfected cells, the antigenome RNA is encapsidated by N to form an antigenome RNP. After the first round of replication, which is achieved by the expressed support proteins, genome RNP is available from which all viral mRNAs can be transcribed, initiating an infectious cycle. The key element for success was the expression of positive-strand antigenome RNA. In contrast to genome-sense RNA, antigenome RNA can not hybridize to the positive-strand mRNAs encoding the viral support proteins, and thus does not interfere with the initial assembly of RNPs.

This method was also applicable to recovery of recombinant VSV rhabdovirus (Lawson et al. 1995; Whelan et al. 1995) and, moreover, of viruses from the other NNSV families, including Sendai virus, measles virus, respiratory syncytial virus or Ebola virus, illustrating the common principle of *Mononegavirales* gene expression (for review see Conzelmann 1998). Variations introduced since then in the NNSV rescue systems mostly pertain to the manner of expression of the plasmid-encoded viral components. As a source of T7 RNA polymerase, different recombinant poxviruses have been used, or nonviral systems including transient expression from transfected plasmids or stable cell lines such as the widely used BSR T7/5 cells (Buchholz et al. 1999; Fuerst et al. 1986). For recovery of the segmented influenza virus, mostly RNA polymerase I (PolI)-driven constructs have been used (Schickli et al. 2001; Neumann et al. 2002). However, PolII has also been found to be suitable for recovery of rabies virus in spite of cytoplasmic replication (Inoue et al. 2003). A comprehensive comparison of the approaches used in the different NNSV systems was published recently (Conzelmann 2004).

Table 1 Reverse genetics systems for rhabdoviruses

Species	Parental strain	Expression system	Reference
RV	SAD B19 (attenuated)	vTF7–3	Schnell et al. 1994; Finke and Conzelmann 1999
	SAD B19 (attenuated)	BSR-T7/5 cells	Finke and Conzelmann 1999
	RC-HL (attenuated)	vTF7–3	Ito et al. 2001
	HEP-Flury (attenuate)	RNA Pol II	Inoue et al. 2003
VSV	Indiana serotype	vTF7–3	Lawson et al. 1995
	Indiana serotype	vTF7–3	Whelan et al. 1995
	Indiana serotype	BSR T7/5 cells	Harty et al. 2001
IHNV		vTF7–3	Biacchesi et al. 2000
SHRV		vTF7–3	Johnson et al. 2000

IHNV, infectious hematopoietic necrosis virus; SHRV, snakehead rhabdovirus.

Of the *Rhabdoviridae,* so far, reverse genetics systems allowing recovery of infectious virus has been achieved for rabies virus, including three different rabies virus strains, for the arthropod-borne VSV, and for the two fish rhabdoviruses: infectious hematopoietic necrosis virus (IHNV) and snakehead rhabdovirus (SHRV) of the novirhabdovirus genus (Table 1).

2
Genetic Engineering of Rhabdoviruses

With efficient reverse genetics systems for rhabdoviruses and other NSVs available, the high degree of amenability of these viruses to genetic engineering and their ability to express foreign genes were rapidly verified. Indeed, the initial report on recovery of recombinant rabies viruses from cDNA included the description of a rabies virus expressing an additional, artificial mRNA from the recombinant genome (Schnell et al. 1994). The potential to express foreign genes was obvious and predicted from the modular genome organization. Since rhabdovirus genomes represent a succession of individual cistrons defined by conserved gene start and gene end signals, and since rhabdoviruses with more than the five basic genes were known, the introduction of additional, and maybe multiple genes appeared easy. This is in contrast to positive-strand RNA viruses such as poliovirus or hepatitis C virus whose genome is translated into a polyprotein-precursor, which is cleaved by proteases to yield the mature gene products. There, specific strategies for expression of foreign genes have to be applied. The potential of rhabdoviruses to serve as useful gene delivery vectors for many applications is supported by several more intrinsic features of this virus group, while others appear to argue against certain applications. Of advantage for gene expression is certainly the helical structure of RNPs, which suggests the lack of strictly defined size limits as it applies to viruses with icosahedral capsids. Then the tight encapsidation of RNA in the RNP is unique for RNA viruses. This is assumed to dramatically reduce the chance of recombination between viruses and of rapid loss of nonessential sequences. Entry of viral RNA or RNPs of mammalian rhabdoviruses into the nucleus of host cells is unlikely, and moreover, the RNP structure may restrict access of reverse transcriptases to the RNA. Therefore, accidental generation of DNA and its introduction into the host cell genome is very unlikely to happen. Finally, many rhabdoviruses, including VSV and RV infect almost all mammalian cell types in cell culture. Infection, gene expression and replication appear to be independent of the cell cycle, such that dividing and nondividing cells can be infected equally well.

Whereas the above features are favorable to vector applications for short-term or transient gene expression, which may include vaccination and cytolytics approaches, it is clear that rhabdoviruses a priori do not meet a major requirement for classical gene therapy, namely long-term expression of transgenes. Usually rhabdovirus-infected host cells are bound to die within a range of hours to days or weeks after infection. Persistent infections have been described for several rhabdoviruses, including IHNV (Kim et al. 1999; Drolet et al. 1995), VSV (Letchworth et al. 1996; Barrera and Letchworth 1996), and rabies virus (Sodja 1980). However, persistence is rarely observed, mechanisms underlying persistence are not understood, and methods to establish and control persistence are not available.

As described in the following, there are manifold immediate technical possibilities to design novel rhabdovirus vectors. In view of these amazing possibilities, it is important to use the new reverse genetics systems also for catching up with knowledge on the biological features of these viruses, in particular with respect to virulence and biosafety.

2.1
Plasticity of Rhabdovirus Gene Expression

Full exploitation of the possibilities provided by rhabdoviruses for transgene expression requires a detailed understanding of the mechanisms of virus gene expression and replication. The genes of rhabdoviruses are separated by conserved regulatory regions (gene borders) comprising a gene end (i.e., transcription stop and polyadenylation) signal and a gene start (i.e., transcription start) signal. The introduction of such a gene border into the genome of recombinant rabies virus, downstream of the G open reading frame in the nontranslated region of the G gene, resulted in transcription of a truncated G mRNA, and an extra short RNA comprising the remainder of the G nontranslated region (Schnell et al. 1994). Expression of a protein-encoding gene merely requires introduction of an open reading frame downstream of the novel gene border. The introduction of a cassette comprising the artificial gene between the G and L genes of RV and of VSV led to stable expression of the reporter gene CAT from the recombinant viruses (Mebatsion et al. 1996b; Schnell et al. 1996b), demonstrating that rhabdoviruses generally are suitable as vectors for transgene expression (Fig. 2A).

It has long been known from studies on VSV that rhabdovirus mRNAs are sequentially transcribed starting from the 3′ terminal genome promoter (Abraham and Banerjee 1976; Ball and White 1976). Due to dissociation of the polymerase at each gene border, a progressive loss towards the 5′ end is observed. This results in a gradient of transcripts following the gene order (Iverson and Rose 1981). Notably, the gene order of natural rhabdoviruses is conserved with the N and P genes needed in stoichiometric amounts for RNP formation, at the first two 3′ proximal positions, whereas the catalytic L protein is encoded by the most 5′ terminally located gene (Conzelmann 1998; Pringle 1997). Due to the transcript gradient, N mRNAs are the most abundant and L mRNAs the least abundant viral mRNAs. This unique feature of *Mononegavirales* gene expression can modulate the level of expression of a transgene by changing the relative distance from the 3′ promoter (Wertz et al. 2002). However, it must be noted that introduction of extra cistron should attenuate transcription of downstream genes, in particular the L gene. This may decrease the fitness of the virus by limiting polymerase. Although expression of one or two extra genes located between the viral G and L genes did not greatly affect replication of recombinant RV or VSV (Mebatsion et al. 1996b; Schnell et al. 1996b; Haglund et al. 2000), insertion of multiple genes inevitably will have attenuating effects. Though introduction of genes at any position is in principle possible, even downstream of L (Finke et al. 2003) or upstream of P (McGettigan et al. 2003b), some may have more pronounced effects of virus replication. A systematic analysis of recombinant VSV revealed that an additional transcription unit between the N and P genes attenuates virus replication leading to a tenfold drop in virus titers (Wertz et al. 2002).

The functions of VSV transcription signals has been extensively examined (Stillman and Whitt 1998, 1999; Barr and Wertz 2001; Barr et al. 1997a, 1997b; Hinzman et al. 2002). These studies showed that each step in transcriptase

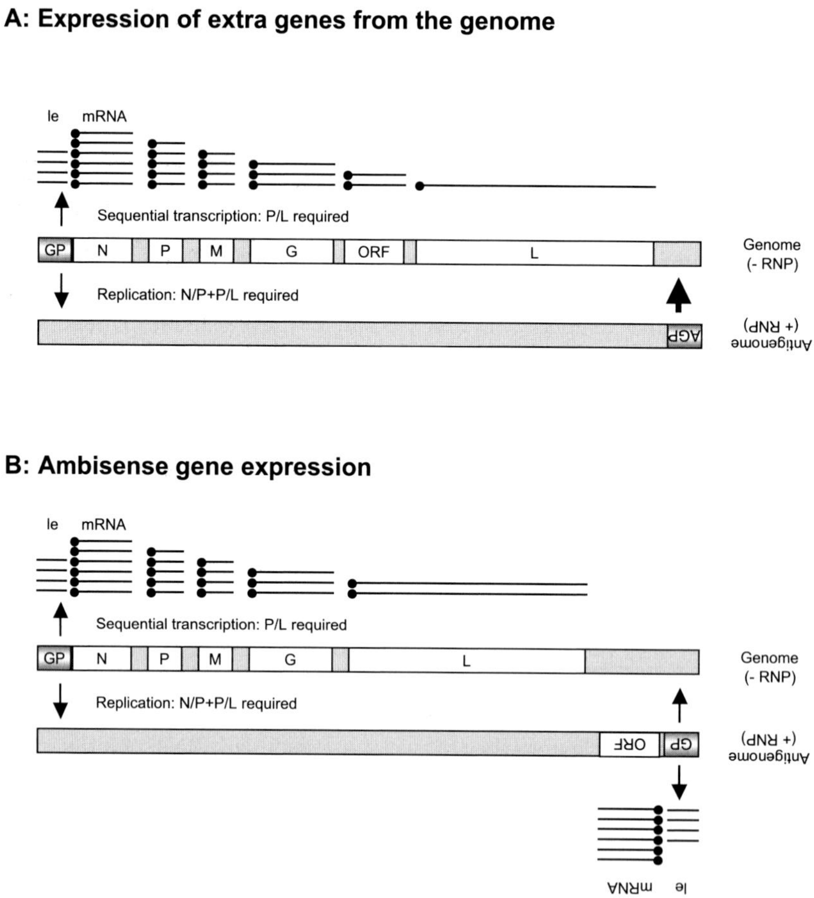

Fig. 2A, B Expression strategies for rhabdovirus vectors. **A** Expression of additional genes from the negative-strand genome. Additional ORFs flanked by extra transcription stop and start signals can be inserted in the virus genome. Since the virus genes are transcribed sequentially, expression of the extra gene is coupled to the transcription of the viral genes and the expression level is determined by the position on the viral genome. **B** In case of rabies virus, replacement of the 3′ end of the antigenome RNA comprising the antigenome promoter (*AGP*) with a copy of the transcriptionally active genome promoter (*GP*) allows expression of genes from the antigenome RNP. In such ambisense vectors, the expression of extra ORFs does not affect the transcription gradient of the viral genes

regulation, including transcription stop, polyadenylation of the transcript and reinitiation of transcription at the re-start signal can be affected by specific mutations, insertions or deletions. These transcription signals are in principle a target for adjusting transgene expression. Notably, the transcription of all VSV genes is driven by highly similar transcription signals, each containing a 2-nt nontranscribed intergenic region (IGR) (Rose 1980). This uniformity and the observation in the above-mentioned mutagenesis studies that no transcription up-mutations were observed suggests that VSV transcription is already preset to maximum. So it appears that VSV transcription can only be downregulated. This may be achieved by changing transcription signals to increase polymerase read-through at the gene borders or to decrease the efficiency of reinitiation.

In contrast to VSV, the four gene borders of RV have nontranscribed IGRs comprising different numbers of nucleotides, namely 2, 5, 5, and 24–29 at the N/P, P/M, M/G, and G/L borders, respectively (Conzelmann et al. 1990; Tordo et al. 1986), indicating the possibility of differential regulation of transcription. Indeed, when the individual IGRs were introduced into the N/P gene border, an attenuation of downstream transcription to 78% (P/M), 81% (M/G) or 11% (G/L) as compared to the authentic N/P gene border was observed. Moreover, in a recombinant RV the replacement of the complete suboptimal G/L gene border with a copy of the N/P border resulted in overexpression of L mRNA, and a general increase in viral RNA synthesis by the increased polymerase concentration (Finke et al. 2000). This showed that RV has evolved mechanisms to modify the steepness of the transcript gradient by further downregulating the expression of distal genes. Thus, at least in case of RV-like rhabdoviruses, unwanted attenuating effects of additional transcription may be overcome by artificially boosting transcription of the L gene.

A quite attractive way for transgene expression from RV-like viruses is by mimicking the expression strategy of a group of segmented NSV, namely arenaviruses, which have so-called ambisense genomes (Fig. 2B). This implies that the rhabdovirus antigenome RNA is modified in a way to allow not only replication but also transcription and requires manipulation of the viral promoters. The 3′ end of the genome and antigenome RNA of rhabdoviruses constitute the promoters for entry of the polymerase (referred to as genome promoter [GP] and antigenome promoter [AGP], respectively). Both drive replication of the RNA, but only the GP is able to drive transcription as well. An important region for transcription is the leader/N gene junction at which probably the decision between replication and transcription is made. The expression of a CAT reporter gene from the antigenome of rRV was achieved by replacing the viral AGP with a copy of the GP including the leader/N gene junction, followed by the CAT gene. Transcription of the antigenome CAT gene was terminated by a modified RV gene junction able to mediate transcription stop and polyadenylation but not reinitiation of downstream transcripts. The promoter exchange not only resulted in efficient CAT expression from the antisense virus but also in accumulation of equal amounts of genome and antigenome RNAs in the infected cells, confirming the idea that the observed overproduction of genome in natural viruses is determined exclusively by the strength of the promoters in supporting replication. Moreover, equal amounts of virus particles with positive- and negative-strand RNA were formed, demonstrating indiscriminate incorporation of RNPs into virus particles independent of putative packaging signals in the RNA (Finke and Conzelmann 1997). Such ambisense viruses appear to be particularly well suited for expression of multiple foreign genes, while leaving the natural N-P-M-G-L gene order and expression of these housekeeping genes from the genome untouched. Surprisingly, the ambisense strategy seems not to be applicable to VSV vectors. In this case, no infectious virions were produced from ambisense RNPs, suggesting the presence of specific packaging signals in the 5′ terminus of genome RNA (Whelan and Wertz 1999). In contrast, recombinant ambisense Sendai viruses perfectly matched the results obtained with RV (Le Mercier et al. 2002), suggesting that this strategy may be applicable to a wider spectrum of NSV vectors.

Ambisense rhabdoviruses not only allow expression of genes from the antigenome, they also are favorable as helper viruses for supporting replication of defective RNAs. Since the relative activity of the terminal replication promoters is mainly determined by competition between the AGP and the weaker GP (Finke and Conzelmann 1997), ambisense RVs containing only GP sequences allow selective amplification and expression of minigenomes comprising a copy of the strong AGP. Wild-type RV is a less altruistic helper virus, as it competes with the minigenomes for replication (Finke and Conzelmann 1999). Since such helper virus systems represent bipartite settings in which the genome encoding the essential virus proteins is physically separate from the genome encoding the gene of interest, this system may be less valuable in in vivo applications. In cell culture, however, the bipartite system may allow effective expression of genes from defective (interfering) genomes. Since the NSV rescue efficiency decreases with increasing length of the full-length cDNA, this might be the method of choice to separately encapsidate and express very large RNAs.

Gene expression of rhabdoviruses cannot only be manipulated by modification of the *cis*-acting internal transcription signals or terminal promoter sequences. Increasing knowledge on the factors that determine the balance of virus transcription and RNP-replication allows fine-tuning of transcription and vector replication by manipulating these factors. The RV matrix protein M, well-known for its importance in virus assembly and budding, has recently been identified as a regulator of RV RNA synthesis (Finke et al. 2003). In the absence of M protein, RV RNA synthesis is shifted towards transcription and this shift is reversible by addition of M protein *in trans*. A recombinant RV, in which M expression was attenuated by moving the M gene to the 5′ end of the genome, transcribed much more mRNA than wt RV but yielded similar infectious titers. Such high-transcription RV vectors can also be created without modifying the gene order or transcription signals, by using M mutants that have lost the ability to regulate transcription and replication but have retained the functions in virus assembly (Finke and Conzelmann 2003).

The limits of the coding capacity of rhabdoviruses have yet to be determined. So far, insertions of up to 4.5 kb into VSV have been reported (Haglund et al. 2000). The helical structure of RNPs does not impose restrictions, rather the techniques available for rescue of recombinant NNSV from cDNA may be limiting. With the present systems, rescue efficiency decreases markedly with increasing length of antigenome RNAs to be encapsidated into functional RNPs. In addition, transcriptional attenuation at the gene borders (Iverson and Rose 1981) may limit the number of additional cistrons, even using improved gene borders.

2.2 Stability of Rhabdoviruses

As compared to other RNA viruses, genetic material introduced into rhabdoviruses is surprisingly stable. Expression of reporter-genes that do not affect virus replication was stable over 25 and 15 virus passages for recombinant RV and VSV, respectively (Mebatsion et al. 1996b; Schnell et al. 1996b). The long maintenance of dispensable sequences may be explained by the poor recombination rate of rhabdoviruses, which is most likely due to the protection of the RNA in the RNP complex. However, after manipulations interfering with virus amplification revertants rapidly appear. The truncation of the VSV G-protein cytoplasmic tail to a single nucleotide resulted in a tenfold reduced incorporation of the protein into virions. After 19 virus passages, a revertant virus was isolated in which the cytoplasmic tails was elongated to eight amino acids by two nucleotide exchanges, restoring wt replication of the virus (Schnell et al. 1998). A frameshift mutation abolishing CD4 expression from recombinant VSV was observed after 26 virus passages. Expression of measles virus F protein, which led to poor growth of recombinant VSV, was also rapidly silenced by mutations in the upstream transcription termination site from 3′-AUAC-5′ to 3′-AUAU-5′ as well as by elongation of the subsequent oligo-U tract. This led to failure of the transcription signal such that a bicistronic VSVG-F mRNA was produced from which only G protein can be translated (Quinones-Kochs et al. 2001; Schnell et al. 1996a). Mutations in the transcription stop-signal leading to transcription of a bicistronic mRNA were also observed after the insertion of an extra transcription unit between the VSV N and P genes. Poor replication of the original recombinant virus was restored already after two virus passages (Wertz et al. 2002). Obviously, silencing gene expression from VSV is predominantly achieved by mutations in the transcription signals, which led to the assumption that these sequences on the viral genome are hot spots for mutations (Quinones-Kochs et al. 2001). However, in all cases analyzed so far, the read-through mutations resulted in a fitness gain of the virus and may have been selected only by selection pressure. Especially the stability of extra transcription units in the more downstream gene borders, that do not affect replication of the virus, argue against the hot-spot theory (Wertz et al. 2002).

Larger deletions in rhabdovirus vectors have not been described so far, misleadingly suggesting that rhabdoviruses are not able to excise sequences by recombination. A known mechanism to delete parts of the genome is the generation of defective interfering particles (DIs). DIs are believed to be generated when the polymerase together with the attached nascent RNA chain dissociates from its template and resumes RNA synthesis either on another site on the same genome or on another genome (Rose and Whitt 2001). The frequent occurrence of DIs after virus passages at high multiplicity of infection supports the idea that NSV genomes can undergo recombination events. The high stability of the natural rhabdovirus genomes is only due to the low frequency of RNA recombination within the NSV group and the low chance for combinations allowing outgrowth of recombinants. Therefore, the immobilization of an ancestral gene order, which has been suggested to be a result of inability to recombine (Ball et al. 1999), is more likely a result of low replicative advantage of such viruses. Indeed, there are examples of homologous and nonhomologous recombination of segmented NSVs such as influenza virus (Bergmann et al. 1992; Khatchikian et al. 1989; Orlich et al. 1994) and Tula hantavirus (Plyusnin et al. 2002). Moreover, very recently genetic recombination between two different recombinant human respiratory syncytial viruses has been demonstrated (Spann et al. 2003).

2.3
Envelope Switching: Flexibility of Rhabdovirus Envelopes

The propensity of rhabdoviruses to accept changes in the composition of the envelope is of great potential for manifold vector applications. Rhabdoviruses carrying novel proteins in their envelopes may have a role to play as particulate vaccines, and as targetable gene delivery vectors (Table 2). Early experiments with ts mutants of VSV (Schnitzer et al. 1979) initially challenged the importance of the rhabdovirus G protein for virus particle formation. The generation of recombinant RV lacking the G gene then confirmed that rhabdovirus particle formation does not require G, though in the presence of G, budding efficiency is increased approximately 30-fold (Mebatsion et al. 1996a). The major driving force for rhabdovirus budding is thus the M protein, as was also confirmed by analysis of a rabies virus M gene deletion mutant (Mebatsion et al. 1999; Finke and Conzelmann 2003). Moreover, it appears that the M proteins of rhabdoviruses alone, when expressed in the absence of any other viral protein, are able to form membrane vesicles released from the cell surface, as demonstrated for VSV M (Li et al. 1993; Sakaguchi et al. 1999) and indicated by microscopy for RV (S. Finke, unpublished results). In this respect, the M proteins of rhabdoviruses share functional properties with retroviral gag-proteins (Garoff et al. 1998). Since, similar to retroviruses, the surface glycoprotein protein is not required for rhabdovirus particle formation, the possibility to exchange G with other glycoproteins, or to co-incorporate them with G into the virus envelope, was intensely investigated.

As the deletion of the G gene renders the recombinant rhabdoviruses defective for infectivity, the protein, or proteins with the analogous functions, must be supplied *in trans* either by stable cell lines or transiently, to produce functionally the so-called pseudotype virions. Alternatively, genes whose products substitute for the functions of the G, for example, can be introduced into the G-deficient viral genome to create so-called surrogate viruses. The transiently complemented pseudotype viruses represent useful and very safe tools, as they allow one round of entry and replication but not production of infectious progeny virus, whereas surrogate viruses are nondeficient and can spread further. Both pseudotype and surrogate viruses have been successfully used to incorporate heterologous glycoproteins into the envelope of G-deficient recombinant rhabdoviruses.

The first example of incorporation of another virus surface protein and of retargeting to a special cell type was a G-deleted RV that was complemented *in trans* with chimeric HIV-1 gp160 protein, comprising the cytoplasmic tail of RV G (Mebatsion and Conzelmann 1996). The pseudotyped virus had gained the ability to specifically infect CD4-positive cells, showing that the heterolo-

Table 2 Vaccination and retargeting vectors

VSV G [2A]	RV	Foley et al. 2000
Second copy of RV G[a]	RV	Faber et al. 2002
G/GFP fusion protein[a]	VSV	Dalton and Rose 2001
VHSV G[b]	IHNV	Biacchesi et al. 2002
Measles virus F/H[a]	VSV	Schnell et al. 1996a
Measles F/H [3B,3]	VSV	Tatsuo et al. 2000
Ebola virus G [2B,3]	VSV	Takada et al. 1997
Ebola virus G [2A,3]	VSV	Takada et al. 2003
HRSV F or G [1]	VSV	Kahn et al. 1999
HRSV F or G [2A]	VSV	Kahn et al. 2001
Hantaan or Seoul virus G [2B,3]	VSV	Ogino et al. 2003
HTLV-1 env[c,d]	VSV	Okuma et al. 2001; Okuma et al. 2003
HIV-1 gp160[c]	RV	Mebatsion and Conzelmann 1996
HIV-1 gp160[a]	RV, VSV	Schnell et al. 2000; Johnson et al. 1997
HIV-1 gp160[a and b,d]	RV	Foley et al. 2002
HIV-1 gp160[b,d] and GFP a	VSV	Boritz et al. 1999
HIV-1 gp160[a] and VSV G exchange[b]	VSV	Rose et al. 2000
HIV-1 Gag[a]	RV	McGettigan et al. 2001b
HIV-1 Gag/gp160[a]	VSV	Haglund et al. 2000
SHIV Gag/Env[a] and VSV G exchange[b]	VSV	Rose et al. 2001
HIV-1 Gag-Pol[a] and HIV-$1_{89.6P}$ Env[a]	RV	McGettigan et al. 2003a
CD4, CXCR4[c,d]	RV	Mebatsion et al. 1997
CD4, CXCR4[b,d]	VSV	Schnell et al. 1997
CD4[a]	VSV	Schnell et al. 1996a
HCV E1E2p7; HCV E2[a]	RV	Siler et al. 2002
HCV E1 or E2[a(ts045),d]	VSV	Lagging et al. 1998
HCV E1[c,d]	VSV	Matsuura et al. 2001
HCV E1 or E2[a and b], E1 and E2[b]	VSV	Buonocore et al. 2002
HCV C/E1/E2[a]	VSV	Ezelle et al. 2002
Sindbis E1 and E2[b and c,d]	VSV	Bergman et al. 2003
BVDV E2[a]	VSV	Grigera et al. 2000
CRPV-L1	VSV	Reuter et al. 2002

[a] Additional gene (together with authentic G).
[b] Replacement of authentic G by insertion into the genome.
[c] *In trans* complementation of ΔG virus.
[d] Successful retargeting.

gous surface glycoprotein was not only incorporated into the virions but was active in receptor binding and subsequent membrane fusion. HIV-1 gp160 surrogate VSV and RV able to amplify only on cells expressing the HIV-1 receptor confirmed the potential of rhabdoviruses in retargeting (Johnson et al. 1997; Foley et al. 2002). An increasing number of VSV and RV pseudotype and surrogate viruses carrying glycoproteins from a variety of viruses, including, for example, Ebola virus or hepatitis C virus (HCV), have been used in the following for purposes of retargeting and for presentation of antigen on the surface of virus particles (Table 2).

One startling example illustrating the great potential of rhabdovirus retargeting using nonviral cellular proteins was provided by rhabdoviruses engineered to target another pathogen. Both in RV and VSV particles, a functional cellular HIV-1 receptor complex consisting of the type I transmembrane protein CD4 and the seven-transmembrane chemokine receptor CXCR4 was successfully incorporated. The resulting virions were not able to enter cells unless they expressed the HIV-1 Env protein or were infected by HIV-1 (Mebatsion et al. 1997; Schnell et al. 1997). In this case, the virus lacked any protein with membrane fusion activity and entry by membrane fusion was mediated by the cell-bound HIV-1 Env protein (see also Sect. 3).

The requirements for incorporation of foreign proteins into rhabdovirus envelopes have not been definitely characterized but they seem to vary for different viruses. Rhabdovirus particle budding takes place predominantly at the cell surface membrane. Surface expression of the foreign membrane proteins is therefore necessary. For efficient incorporation into RV particles, the cytoplasmic tail of the heterologous transmembrane protein must be replaced with that from the RV G tail, or at least with that from related homotypic G proteins, like those from other Lyssaviruses, such as Mokola virus (Mebatsion et al. 1995). This suggests rather specific interactions of the cytoplasmic G with the M protein layer beneath the virus membrane (Mebatsion et al. 1996a). In contrast, for uptake by VSV particles, a correct length of the cytoplasmic tail is more important than specific amino acid sequences (Schnell et al. 1996a, 1998). The reason for these differences may lie in the structure of the respective M layer, which may provide a lattice for accommodation of G protein. It should also be considered that assembly of G protein with the M layer may modify its intrinsic membrane bending and budding efficacy. As outlined above, particles are formed in the absence of G, but its presence results in an approximately 30-fold increase in virion yield. The different heterologous chimeric proteins may fulfill the functions of the authentic G in supporting M budding more or less well. Indeed, the membrane-proximal amino-acids of the VSV G ectodomain have been shown to promote particle formation, since G proteins with deletions in this region produce 10- to 20-fold fewer virions (Robison and Whitt 2000). Moreover, the stem region (G stem) containing these membrane-proximal amino acids were reported to improve the membrane fusion activity of co-incorporated heterologous viral fusion proteins (Jeetendra et al. 2002).

3 Applications

3.1 Attenuation of Rhabdoviruses

Both VSV and RV are long-known pathogens. The rhabdovirus reverse genetic systems now allow study of virus pathogenicity and virulence in more detail, which is expected to lead to the generation of viruses and virus vectors that are acceptable for live applications. The approaches taken include transfer of attenuating mutations identified by classical means to recombinant viruses and their combination in a single virus backbone. In addition, empirical and hypothesis-driven approaches may reveal the identity of virulence factors and their elimination. In particular, the potential neurotropism of a rhabdovirus vector is a matter of concern.

VSV causes a highly contagious disease in horses, cattle and pigs, characterized by severe lesions in the oral mucosa and udder. Intranasal and intracerebral inoculation of wild-type VSV into mice causes fatal encephalitis (Sabin and Olitsky 1938; Miyoshi et al. 1971). However, in spite of the ability of VSV to infect the central nervous system, so far only a single case of human encephalitis associated with VSV is known (Quiroz et al. 1988). Rabies virus is a prototype neurotropic virus that enters the central nervous system (CNS) via retrograde axonal transport. Virus establishment in the brain causes fatal disease in humans and animals. Safe live RV vaccines based on attenuated viruses such as the SAD (Street Alabama Dufferin) B19 are available and have been used for decades for wildlife immunization. Nevertheless, the plethora of potential rhabdovirus vector applications in vivo is an issue raising concerns, especially with regard to the potential neuropathogenicity of rhabdovirus vectors.

The major neuropathogenicity factor of VSV and RV is the surface glycoprotein G (Morimoto et al. 2000; Martinez et al. 2003), which allows more or less specific entry into neuronal cells. Attenuating mutations in the RV G protein have been identified. Mutation of amino acid position 333, which is located in a region referred to as antigenic site III of RV G protein, decelerates virus uptake by the cell (Dietzschold et al. 1983), and infection of motoneurons and virus spread in neuroblastoma cell cultures is hampered (Coulon et al. 1998). In adult mice, a highly reduced neuroinvasivity is observed. However, mutations at G position 333 do not result in virus attenuation in all RV strains (Morimoto et al. 2001a). A residual pathogenicity in immune-immature 1- to 2-day-old suckling mice is always observed, thus indicating a potential risk for immune-deficient or immune-compromised individuals.

One reverse genetics approach described to generally attenuate rhabdoviruses by modification of the G is the truncation of the cytoplasmic tail of the G proteins. However, infectious titers of recombinant RVs harboring tail-less or truncated G in cell cultures were six- to tenfold decreased (Morimoto et al. 2001a; McGettigan et al. 2003b; Mebatsion et al. 1996a). Recombinant VSVs in which only one amino acid of the cytoplasmic domain of G was retained replicated to titers of 10^7 PFU/ml, which is 100-fold below normal wt-virus titers. Moreover, rapid emergence of revertants after a couple of passages in cell culture has been described (Schnell et al. 1998). Nevertheless, such attenuated recombinant VSVs have been used successfully to elicit immune responses to heterologous epitopes expressed from the virus (Roberts et al. 1999), although the ability to induce antibody response decreased with increasing tail attenuation (Schlereth et al. 2003).

Although the glycoprotein is a suitable target for attenuation of individual virus strains or isolates, it is not the prime target for attenuating rhabdovirus vectors in general. Attenuating mutations in the glycoprotein are mostly specific and cannot be easily transferred to G proteins of other virus isolates (Morimoto et al. 2001a) or other rhabdovirus groups. Obviously, in the retargeting or vaccination regimens involving substitution of the G protein with other surface proteins (see Table 2), the tropism exclusively depends on the foreign protein. Though most replacements reported so far resulted in reduced glycoprotein incorporation, stability, infectivity, and spread of the recombinant virus, the new combinations of envelope proteins with RNPs, which are able to replicate in virtually all cell types, require attention.

Other factors that influence pathogenicity of rhabdoviruses are known. The attenuation of different RV isolates seems to correlate with their ability to induce apoptosis in infected primary neurons (Morimoto et al. 1999). By overexpression of pro-apoptotic cytochrome C or merely by (over)expression of the viral glycoprotein from an extra gene in recombinant RV, the mortality in mice was reduced. This correlated with significantly higher virus-neutralizing antibody (VNA) titers. The attenuation of these viruses is thought to be due to the release of endogenous adjuvants that stimulate cytotoxic T-cell responses and to improved maturation of dendritic cells and antigen presentation (Pulmanausahakul et al. 2001; Faber et al. 2002).

Another possibility to attenuate rhabdoviruses is to make them more sensitive to the innate immune system. Rhabdoviruses are sensitive to interferons, which means that virus replication is heavily affected by IFN-stimulated gene products. Interestingly, VSV appears to be variable in IFN-induction (Marcus et al. 1998). The lack of IFN induction by natural viruses has been attributed to the activity of the matrix protein M as several mutations in the M gene caused potent IFN-induction (Ferran and Lucas-Lenard 1997). A direct correlation of the ability of M to cause a general host-cell transcription shut-off and the ability to prevent IFN-production was revealed recently (Ahmed et al. 2003). Any mutation in M affecting host-cell transcription shut off and thereby allowing IFN to be produced should be detrimental for VSV in vivo. The replication of such viruses should be restricted to IFN-incompetent tissue, as is the case for many tumors (see Sect. 3.4).

The RV phosphoprotein has been suggested to be a target for virus attenuation because of its ability to bind dynein light chain 1 (DLC1) (Raux et al. 2000; Jacob et al. 2000). DLC-1-binding could be relevant for the retrograde axonal transport of RV to the CNS. Mutations in the DLC1 binding motif did not affect the functions of P as a cofactor for the polymerase in virus transcription (Poisson et al. 2001), suggesting the possibility of specific attenuation of RV transport. In vivo attenuation of corresponding recombinant RVs, however, were only moderately attenuated in mice, only in combination with the above-described G antigenic site III mutation was a reasonable attenuation in suckling mice achieved (Mebatsion 2001).

A generally applicable method to attenuate VSV, and most likely other rhabdoviruses as well, is based on changing the relative expression levels of the viral proteins by rearrangement of the gene order (Wertz et al. 1998). In particular the translocation of the promoter-proximal N gene to more downstream positions is effective in attenuation. With increasing distance to the 3′ terminal promoter, a stepwise reduction of N mRNA transcription, N protein accumulation, virus replication, and pathogenicity in mice was observed (Wertz et al. 1998). Recombinant VSV in which the N gene was moved to the fourth position and which had the G gene at the first or third position on the genome were nonpathogenic in swine and protected against challenge with wild-type VSV (Flanagan et al. 2001). Moreover, these viruses were cleared from the brain of mice by 7 days after inoculation (Flanagan et al. 2003). Selective gene attenuation by rearrangement of the genome is thus a powerful tool for attenuation of rhabdoviruses. As revertants able to compensate the severe rearrangements are very unlikely to emerge, this method is considered to stable attenuate the viruses. Reducing the expression of the essential P protein is also effective in attenuating viruses. A RV in which an extra transcription unit was introduced between the N and P gene was nonpathogenic for mice even after i.c. injection (McGettigan et al. 2003b).

A safe means to attenuate viruses is to delete essential virus genes and to produce progeny virions in packaging cells providing the lacking gene product. Complementation *in trans* of rhabdoviruses was first employed for pseudotyping G gene-deleted RV and VSV with their own or heterologous surface glycoproteins (Mebatsion et al. 1996a; Takada et al. 1997). G gene-deficient RV (RVΔG) pseudotyped with the RV G (RVΔG+G) is nonpathogenic. After stereotaxic inoculation into the striatum of rats, infection was restricted to the initially infected neurons and the virus was not able to enter secondary cells (Etessami et al. 2000). Thus, such viruses represent suitable single-round vectors for certain applications.

Another RV gene that can be deleted to avoid uncontrolled virus spread is the M gene. In the absence of M, the formation and budding of extracellular progeny virus is abolished. In some cell cultures, formation of syncytia is observed, due to accumulation of RVΔM-encoded G protein. In mice, RVΔM is apathogenic after intracerebral injection (Mebatsion et al. 1999). The deletion of M provides an attractive tool to avoid uncontrolled spreading of surrogate rhabdovirus harboring foreign fusion-active surface glycoproteins.

A pre-condition for reasonable use of such gene deletion vectors is a powerful complementation system. Since both rhabdovirus M and G proteins are cytopathic, the generation of packaging cell lines is a problem. However, the use of tight conditional expression, such as the Tet-on/off system, has allowed the establishment of packaging cells in which, for example, expression of RV M protein, or both M and G, can be induced (Finke et al. 2003). In these systems, production of M and/or G-deleted RV titers of up to 10^7 infectious virus particles per milliliter supernatant (Finke, unpublished results).

3.2 Cell Targeting

The G protein of rhabdoviruses mediates both attachment to cell receptors and membrane fusion. Retargeting of rhabdoviruses is most easily achieved by replacing the G protein with analogous viral proteins combining these two activities as well, such as HIV-1 Env, as described above (Mebatsion and Conzelmann 1996; Schnell et al. 1996a). In most cases the rather broad host cell range of the parental rhabdoviruses is thereby restricted to cells that express the respective receptor. HIV-1 Env pseudotype or surrogate viruses thus enter only cells expressing CD4 and a chemokine receptor or other HIV-co-receptors (Mebatsion and Conzelmann 1996; Johnson et al. 1997; Foley et al. 2002; Schnell et al. 2000). Similarly, G-deficient VSV has been re-targeted with the envelope protein of human T-cell leukemia virus type 1 (HTLV-1). Such HTLV-1-pseudotyped VSVs were used to study effects of different components on HTLV-1 entry (Okuma et al. 2001, 2003).

Also, Ebola virus G protein is alone sufficient to mediate attachment to cells and membrane fusion, as first demonstrated by pseudotyping of a G-deficient VSV (Takada et al. 1997). Whereas virus complemented with authentic VSV G protein revealed quite similar infectivities on a broad range of cell lines, the retargeted virus was 100-fold less infectious on non-primate cells and was thus mimicking the tropism of Ebola virus (Takada et al. 1997). The infectivity of pseudotypes was lower than that of virus complemented with the authentic VSV G protein, as has been observed previously with HIV-1 Env, suggesting some loss in glycoprotein function or rather some impairment on virus formation. Such retargeted viruses do allow the study of Ebola virus tropism and immunogenicity at lower biosafety levels. For instance, by cultivation of a recombinant VSV stably expressing the Ebola virus glycoprotein in the presence of Ebola virus-neutralizing Mabs, escape mutants were isolated, which revealed the identity of epitopes on the glycoprotein (Takada et al. 2003).

Studies employing rhabdovirus pseudotyping to address functions of glycoproteins from dangerous virus, or from viruses that cannot easily be cultivated in vitro, also included hepatitis C virus (HCV). The results of these studies are controversial. First, a temperature-sensitive VSV mutant (tsO45), which does not express functional VSV G protein at a nonpermissive temperature, was used for complementation by either the E1 or E2 protein of HCV. The formation of infectious virus at a nonpermissive temperature was reported for either E protein (Lagging et al. 1998). Both E1 and E2 proteins were reported to be necessary to increase the infectivity above the background level of a G-deficient VSV (Matsuura et al. 2001). Finally, a third group complemented VSVΔG with E1 or E2 alone, or in combination. HCV glycoprotein-dependent infectivity was not detected (Buonocore et al. 2002).

It is indeed possible to substitute for the attachment and membrane fusion functions of the rhabdovirus G by its replacement with two (or more) proteins active in either of these functions, as shown by the infectivity of VSV pseudotyped with the H and F proteins of measles virus. In this approach, the type II glycoprotein H is co-incorporated through binding to F, which, like G, is a type I membrane protein. As may have been predicted, the F protein of measles virus alone was not sufficient for infectivity of pseudotyped VSV and RV (unpublished data), illustrating the requirement of the H and binding to one of the measles virus receptors. The tropism of the resultant viruses was identical to that of the measles virus, revealing that the cell tropism of MV strains is primarily determined by virus entry (Tatsuo et al. 2000).

For entry of rhabdoviruses into target cells, protein(s) active in virus attachment and membrane fusion are required. Most intriguingly, however, such proteins do not have to be located in the viral membrane. This was illustrated first by rhabdoviruses with envelopes carrying complexes of CD4 and CXCR4, or of CD4 and CCR5, i.e., complete HIV-1 receptor complexes (Mebatsion et al. 1997; Schnell et al. 1997). Such viruses are not able to enter natural cells since they lack a protein with membrane fusion activity. Only when they encounter cells to which they can bind appropriately and which themselves provide a membrane fusion activity, can they enter and infect the cell. Indeed, this applies only to cells expressing HIV-1 Env, or HIV-1-infected cells (or so far unknown viruses that might use CD4 and/or one of the chemokine receptors for entry). These viruses represent, therefore, the first example of a virus targeted to cells infected by another pathogen. Indeed, when HIV-1 was co-cultured in a T cell line with such VSV-derived hunter viruses, its replication was reduced up to 10^4-fold, through rapid killing of the HIV-1 infected cells by the superinfecting hunter viruses (Schnell et al. 1997). In such an in-vitro predator–prey scenario and in order not to let escape the fast-growing HIV-1, an even faster growing and highly cytopathic hunter virus such as VSV is required, or highly up-tuned RV, which usually grows slowly and with little cytopathic effect. A mathematical model predicted that HIV-1 load in vivo can be reduced by 92% and that a recovery of host T cells to 17% of their normal level could be achieved (Revilla et al. 2003). However, this model does not consider predicted innate and adaptive immune responses to the anti-HIV virus. Another problem with VSV hunter viruses was the low efficiency of progeny virus production, related to the absence of supporting functions of the authentic viral G protein. Indeed, this could be considerably improved by inclusion of so-called G stem constructs, which represent short VSV G transmembrane protein with only a short stretch of the membrane-proximal ectodomain residues. This G stem effectively supports budding of G-deleted VSV (Robison and Whitt 2000) and of VSV carrying the HIV-2 receptor. This virus was successful in reducing HIV-2 load in macaques with acute and chronic infection (Whitt et al. 2003).

In summary, rhabdovirus retargeting is and will be successful with a series of proteins having attachment and fusion activity combined, or with pairs of cooperate proteins with separate activity. Since there are several possibilities to add novel binding activities to proteins, for example through fusion with known ligands, binding peptides, or single-chain antibody fragments, the challenge in virus retargeting viruses is to achieve attachment-dependent membrane fusion. This is hard to achieve without more knowledge on the mechanistics of viral fusion proteins. Promisingly, a recent approach using an attachment negative mutant of Sindbis virus E2 protein with an immunoglobulin G Fc-binding domain insertion was successful in antibody-directed targeting of recombinant VSV to breast cancer cells (Bergman et al. 2003).

3.3 Rhabdovirus-Vector Vaccines

A prerequisite for viral vector applications is a low seroprevalence within the population. The prevalence of VSV antibodies within the human population is low and is mostly restricted to regions where VSV is enzootic (Cline 1976; Tesh et al. 1969) and to people with a high risk of exposure. Similarly, prevalence of antibodies against RV is also low in the human population and is mostly restricted to vaccinated people at a high risk of exposure or to individuals having received postexposure treatment.

Another criterion is the efficiency of stimulating the immune system. Rhabdoviruses such as RV and VSV are well known to induce strong humoral and cellular immune responses. The VSV G protein is able to activate B cells directly, resulting in a very early generation of virus-neutralizing antibodies, which are most important and sufficient for clearing virus. This immediate and T-helper-cell independent response is thought to be due to the presentation of the VSV G protein in the form of a densely packed paracrystalline layer displaying repetitive epitopes (DeMattos et al. 2001; Zinkernagel 1997). The efficient gene expression and the cytopathic type of infection further leads to strong humoral and cellular immune responses also directed against the other virus proteins, including the RNP.

RV nucleocapsids have been shown to function as an exogenous superantigen specific for Vβ8- and Vβ6 T-cell-receptors (TCR) in humans and mice, respectively (Lafon et al. 1992; Lafon et al. 1994). Superantigens usually bind directly to class II major histocompatibility proteins and elicit a powerful proliferative response of T lymphocytes. In mice bearing the Vβ6 TCR, RV nucleocapsids were able to stimulate both T- and B-cell-specific immune responses against the co-administered influenza virus HA antigene (Astoul et al. 1996), demonstrating the high potential of RV as a vaccine vector with intrinsic adjuvant function.

While the remarkable immunogenicity of rhabdovirus particles is perfect in terms of eliciting initial immune responses, it may preclude multiple use of the same vector for boosting in heterologous vaccine applications. This, however, can be circumvented by using vectors with different envelope proteins, as demonstrated by the group of J. Rose who produced a series of VSV vectors with G proteins from different VSV serotypes that are not cross-reacting (Rose et al. 2000). Similarly, the surface glycoprotein of a RV vector was exchanged

Table 3 Rhabdovirus vectors for the expression of nonviral proteins

Expressed protein	Vector	Reference
CAT[a]	RV, VSV	Mebatsion et al. 1996b Schnell et al. 1996b
GFP[a]	VSV	Boritz et al. 1999
GFP[b] and DsRed[b]	RV (ΔMΔG)	Finke et al. 2003
CAT[b] or GFP[b]	IHNV(ΔNV)	Biacchesi et al. 2000
α RV Mabs[a] + VSV G[b]	RV	Morimoto et al. 2001b
cytochrome C[a]	RV	Pulmanausahakul et al. 2001
TK or IL-4[b]	VSV	Fernandez et al. 2002
IL-2, -4, -12, IFNγ or GM-CSF[b]	VSV (ΔG)	Miller et al. 2004; Klas et al. 2002
CD/UPRT[a]	VSV	Porosnicu et al. 2003
CD4, CXCR4[b]	VSV(ΔG)	Schnell et al. 1997
CD4[a]	VSV	Schnell et al. 1996a

CAT, chloramphenicol acetyl transferase; CD/UPRT, cytosine desaminase/uracil phosphoribsyltransferase; GFP, green fluorescent protein; GM-CSF, granulocyte-macrophage colony-stimulating factor; DsRed Discosoma sp., red protein; Mab, monoclonal antibody; TK, thymidine kinase; IL, interleukin; IFN, interferon.
[a] Additional gene.
[b] Virus genes replaced by heterologous gene.

with that of VSV (Foley et al. 2000), resulting in a virus that is able to replicate in the presence of RV-neutralizing antibodies (Morimoto et al. 2001b).

Because of their remarkable immunogenicity, the potential of recombinant rhabdoviruses to serve as carriers for immunization against a variety of pathogens is being evaluated (Table 2). Most preferably, heterologous antigens are displayed on the surface of rhabdovirus particles to mimic the paracristalline assay of G, and to stimulate generation of antibodies.

A major focus in the field of rhabdovirus-based vaccines is immunization against HIV-1. As it is assumed that an HIV vaccine requires high antibody titers and cytotoxic T lymphocyte response, a live rhabdovirus-based vaccine might have several advantages (Schnell 2001). Both VSV and RV are under evaluation as live vaccines against HIV-1. The HIV-1 gp160 gene was expressed from both viruses and modified such that it was incorporated into the envelope of virus particles. A single infection of mice with a recombinant RV in combination with a single boost using recombinant gp120 elicited a strong humoral response including neutralizing antibodies to the HIV-1 strain from which the antigen was derived (Schnell et al. 2000). Moreover, a solid and long-lasting memory CTL response specific to HIV-1 envelope protein was obtained after a single injection into mice (McGettigan et al. 2001a). A similarly efficient

CD8+ CTL response against HIV-1 Gag was observed after infection with a recombinant RV expressing HIV-1 gag sequences (McGettigan et al. 2001b). High neutralizing antibody titers in an HIV primary isolate could be induced by immunization with an Env-expressing VSV, boosting with corresponding vectors with G proteins from different VSV serotypes (Rose et al. 2000), and therefore escaping neutralization by antibodies induced in the previous infections. Moreover, VSV vectors expressing both HIV Env and Gag led to the formation of HIV-like particles (VLP), which might trigger potent immune response (Haglund et al. 2000). Indeed, strong primary CD8+ T-cell response and robust recall and long-term memory T-cell responses to HIV-Gag and Env could be generated in mice with such VSV vectors (Haglund et al. 2002a, 2002b), further underlining the potential of rhabdovirus vectors as vehicles for the expression of complex antigens. SHIV Gag- and Env-expressing VSV vectors have recently been tested in rhesus monkeys. After challenge with a pathogenic SHIV virus, all monkeys remained healthy and the SHIV load remained low or undetectable (Rose et al. 2001; Ramsburg et al. 2004).

Rhabdoviruses expressing HCV components are also under evaluation for their potential as vaccines. Recombinant VSVs expressing HCV core, E1, and E2 proteins, which results in the formation of HCV-like particles, have been used to immunize mice. Cell-mediated immune response to all of the HCV structural proteins, and humoral response to the E2 protein were detected (Ezelle et al. 2002). Similarly, both types of immune responses were elicited after immunizing mice with recombinant RVs expressing either HCV E2 alone or the HCV E1E2p7 precursor protein (Siler et al. 2002).

A recombinant VSV expressing the influenza virus hemagglutinin (HA) gene was tested for immune response in mice. After intranasal application, VSV-HA raised high titers of neutralizing antibodies to influenza virus and completely protected mice from bronchial pneumonia caused by challenge with a lethal dose of influenza A virus (Kretzschmar et al. 1997; Roberts et al. 1998). A recombinant VSV expressing measles virus hemagglutinin (VSV-H) was used for immunization of cotton rats (Schnell et al. 2000). Even in the presence of passively transferred antibodies to MV H protein, high titers of MV neutralizing antibodies were induced by VSV-H. This suggests that such viruses probably allow induction of neutralizing antibodies even in the presence of maternal antibodies (Schlereth et al. 2000). Respiratory syncytial virus (HRSV) fusion protein (F) and glycoprotein (G) expressed from a recombinant VSV were able to elicit RSV neutralizing antibodies sufficient for protection against HRSV in mice. Interestingly, the induction of protecting antibodies dependent on the presence of VSV-G protein. In the absence of G, HRSV-F antibodies were not neutralizing, and VSV-G antibodies were not induced. This indicated that the degree of the immune response critically depended on the ability of the live vaccine to replicate and spread to a certain degree (Kahn et al. 2001). Also, expression of the major capsid protein (L1) of cottontail rabbit papillomavirus (CRPV) was highly immunogenic when expressed from VSV and protected rabbits from CRPV-induced papillomas (Reuter et al. 2002).

3.4 Oncolytic Viruses

Oncolytic activity of natural negative-strand RNA viruses has been known for 30 years, from experiments published in 1974 towards treatment of human cancers with mumps virus (Asada 1974). More recent oncolytic approaches with RNA viruses (recently reviewed by Russell [2002] and Stanziale and Fong [2003]) involve other natural paramyxoviruses such as Newcastle disease virus (Sinkovics and Horvath 2000; Lorence et al. 1994), measles virus (Grote et al. 2001) or recombinant influenza virus (Bergmann et al. 2001) as well as rhabdoviruses.

VSV is particularly sensitive against the antiviral effects induced by type I interferon (IFN). Many tumors are defective in their response to IFN and thus are not reacting to the growth inhibitory and apoptotic functions of IFN. Although interferon nonresponsive cancer cells may have acquired a survival advantage over their normal counterparts, they may have simultaneously compromised their antiviral response. Indeed, VSV rapidly replicated in and selectively killed a variety of human tumor cell lines even in the presence of doses of interferon that completely protected normal human primary cell cultures. A single intratumoral injection of VSV was effective in reducing the tumor burden of nude mice bearing subcutaneous human melanoma xenografts (Balachandran and Barber 2000; Stojdl et al. 2000). Thus, treatment of IFN nonresponsive tumors with live rhabdoviruses may be a worthwhile strategy. Oncolytic activity of VSV is effective against tumors exhibiting aberrant p53, Ras, or Myc function and involves the induction of apoptosis (Balachandran et al. 2001). The oncolytic activity of wild-type VSV could even be increased by expressing herpes virus thymidine kinase (TK) or interleukin-4 (IL-4) from VSV vectors (Fernandez et al. 2002). Also, recombinant VSV expressing biologically active IFN-β grew to high titers in IFN-incompetent cells and retained oncolytic activity against metastatic lung disease in immunocompetent animals. Paracrine activation of IFN pathways in surrounding cells might represent an additional attenuation factor and antiproliferative effects of IFN might result in stimulation of NK and cytotoxic T cells as well as in dendritic cell (DC) activity, all enhancing antitumor immune responses (Obuchi et al. 2003).

The availability of reverse genetics systems to enhance the intrinsic oncolytic activity of recombinant VSVs and to more specifically target tumors will be of great advantage towards the development of effective and safe oncolytic viruses. Especially viruses modified in their interplay with the potent IFN system will be interesting candidates. Interferon induction is highly variable in VSV isolates and laboratory strains (Marcus et al. 1998) and is correlated with the loss of host-cell shut-off by the viral M protein (Ferran and Lucas-Lenard 1997; Ahmed et al. 2003; Lyles 2000). Viruses that induce IFN and therefore are attenuated in non-tumor tissue while retaining the desired high cytotoxic effects of the M protein are preferable for such applications. Expression of appropriate cytotoxic genes and cytokines will make it possible to further improve cell killing and to modulate the host immune response towards oncolytic activity and tumor-specific immunity.

So far, oncolytic approaches with VSV have been confined to injection of the vector into the tumor in situ. Other strategies may have advantages for certain tumors. Re-injection of human oncolysates prepared in vitro by infection with influenza A virus, NDV or vaccinia virus (Sinkovics 1991; Cassel and Murray 1992; Wallack et al. 1998; Schirrmacher et al. 1998) have been used to activate active tumor-specific immune responses (Sinkovics and Horvath 2000). Recombinant rhaboviruses, expressing the appropriate immunomodulatory cytokines, may offer an alternative system. Moreover, exploitation of the possibility of rhabdovirus retargeting might in the future bring about viruses that are specifically targeted to defined tumor cells and which might be considered for systemic administration in vivo.

3.5 Production of Biologically Active Compounds by Rhabdoviruses

Due to their effective gene expression, rhabdovirus vectors are not only suited for expression of antigen, but also offer a tool for large-scale production of proteins in mammalian cells. The ability of VSV and RV to infect a wide variety of mammalian cell lines and the high sensitivity to UV radiation or chemical treatment, which allows vector inactivation, make these systems a versatile tool for transgene expression. In case of RV, the lack of cytopathic effects on particular cell lines allows long-term production of the gene product over time periods of more than 2 weeks (Prosniak et al. 2003). The efficient gene expression and host-cell shut-down by VSV allows short-term high-level protein expression.

Examples for protein production in rhabdovirus vector systems so far include cytokines and antibodies (Table 3). A VSV vector expressing IFN-β from an extra gene showed largely unaffected growth in cells that are not responsive to IFN and produced high amounts of IFN secreted into the cell culture supernatant (Obuchi et al. 2003). Other adjuvants expressed from recombinant VSV include interleukin-12 (IL-12) which is known to shift the immune response towards the Th1-type. A biologically active, heterodimeric IL-12 fusion protein (IL-12F) consisting of the p40 and p35 chains was efficiently expressed in BHK cells from a G-deficient VSV vector. After 17 h of infection and clarification of vector virus, more than 95% of the protein content of the supernatant was virus-derived vIL-12F. Co-administration of the vIL-12F with a poorly immunogenic listerial antigen elicited a strong cell-mediated immune responses that conferred long-lasting protective immunity (Klas et al. 2002). Expression of interleukin-4 (IL-4) from a recombinant VSV also led to the release of substantial amounts of cytokine into the supernatant of BHK cells. IL-4 is known to drive the immune system to a humoral Th2-type response, making such recombinant viruses also attractive vector candidates for tumor lysis applications. An increased oncolytic activity of the IL-4 expressing virus in an animal tumor model was assumed not to be due to a Th2-type bias but to an increased presence of infiltrating eosinophils and neutrophils (Fernandez et al. 2002).

Finally, a RV has been described that expresses RV neutralizing human monoclonal antibodies. Obviously, this required modification of the RV vector. The anti-rabies immunoglobulins were expressed from a RV in which the targets of the antibody, RV G, was substituted by the VSV G protein (Morimoto et al. 2001b). A cocktail of three human RV-vector derived RIGs, each directed against a different lyssavirus serotype, was successfully tested for the protection of RV-infected mice and hamsters, indicating that such virus-expressed antibodies are fully functional and could be used for postexposure prophylaxis after inactivation of the RV vector by UV-irradiation or chemicals (Prosniak et al. 2003).

References

Abraham G, Banerjee AK (1976) Sequential transcription of the genes of vesicular stomatitis virus. Proc Natl Acad Sci U S A 73:1504–1508

Ahmed M, McKenzie MO, Puckett S, Hojnacki M, Poliquin L, Lyles DS (2003) Ability of the matrix protein of vesicular stomatitis virus to suppress beta interferon gene expression is genetically correlated with the inhibition of host RNA and protein synthesis. J Virol 77:4646–4657

Asada T (1974) Treatment of human cancer with mumps virus. Cancer 34:1907–1928

Astoul E, Lafage M, Lafon M (1996) Rabies superantigen as a Vbeta T-dependent adjuvant. J Exp Med 183:1623–1631

Balachandran S, Barber GN (2000) Vesicular stomatitis virus (VSV) therapy of tumors. IUBMB Life 50:135–138
Balachandran S, Porosnicu M, Barber GN (2001) Oncolytic activity of vesicular stomatitis virus is effective against tumors exhibiting aberrant p53, Ras, or myc function and involves the induction of apoptosis. J Virol 75:3474–3479
Ball LA, White CN (1976) Order of transcription of genes of vesicular stomatitis virus. Proc Natl Acad Sci U S A 73:442–446
Ball LA, Pringle CR, Flanagan B, Perepelitsa VP, Wertz GW (1999) Phenotypic consequences of rearranging the P, M, and G genes of vesicular stomatitis virus. J Virol 73:4705–4712
Barr JN, Wertz GW (2001) Polymerase slippage at vesicular stomatitis virus gene junctions to generate poly(A) is regulated by the upstream 3′-AUAC-5′ tetranucleotide: implications for the mechanism of transcription termination. J Virol 75:6901–6913
Barr JN, Whelan SP, Wertz GW (1997a) cis-Acting signals involved in termination of vesicular stomatitis virus mRNA synthesis include the conserved AUAC and the U7 signal for polyadenylation. J Virol 71:8718–8725
Barr JN, Whelan SP, Wertz GW (1997b) Role of the intergenic dinucleotide in vesicular stomatitis virus RNA transcription. J Virol 71:1794–1801
Barrera JC, Letchworth GJ (1996) Persistence of vesicular stomatitis virus New Jersey RNA in convalescent hamsters. Virology 219:453–464
Bergman I, Whitaker-Dowling P, Gao Y, Griffin JA, Watkins SC (2003) Vesicular stomatitis virus expressing a chimeric Sindbis glycoprotein containing an Fc antibody binding domain targets to Her2/neu overexpressing breast cancer cells. Virology 316:337–347
Bergmann M, Garcia-Sastre A, Palese P (1992) Transfection-mediated recombination of influenza A virus. J Virol 66:7576–7580
Bergmann M, Romirer I, Sachet M, Fleischhacker R, Garcia-Sastre A, Palese P, Wolff K, Pehamberger H, Jakesz R, Muster T (2001) A genetically engineered influenza A virus with ras-dependent oncolytic properties. Cancer Res 61:8188–8193
Biacchesi S, Thoulouze MI, Bearzotti M, Yu YX, Bremont M (2000) Recovery of NV knockout infectious hematopoietic necrosis virus expressing foreign genes. J Virol 74:11247–11253
Biacchesi S, Bearzotti M, Bouguyon E, Bremont M (2002) Heterologous exchanges of the glycoprotein and the matrix protein in a Novirhabdovirus. J Virol 76:2881–2889
Boritz E, Gerlach J, Johnson JE, Rose JK (1999) Replication-competent rhabdoviruses with human immunodeficiency virus type 1 coats and green fluorescent protein: entry by a pH-independent pathway. J Virol 73:6937–6945
Buchholz UJ, Finke S, Conzelmann KK (1999) Generation of bovine respiratory syncytial virus (BRSV) from cDNA: BRSV NS2 is not essential for virus replication in tissue culture, and the human RSV leader region acts as a functional BRSV genome promoter. J Virol 73:251–259
Buonocore L, Blight KJ, Rice CM, Rose JK (2002) Characterization of vesicular stomatitis virus recombinants that express and incorporate high levels of hepatitis C virus glycoproteins. J Virol 76:6865–6872
Cassel WA, Murray DR (1992) A ten-year follow-up on stage II malignant melanoma patients treated postsurgically with Newcastle disease virus oncolysate. Med Oncol Tumor Pharmacother 9:169–171
Cline BL (1976) Ecological associations of vesicular stomatitis virus in rural Central America and Panama. Am J Trop Med Hyg 25:875–883
Conzelmann KK (1998) Nonsegmented negative-strand RNA viruses: genetics and manipulation of viral genomes. Annu Rev Genet 32:123–162

Conzelmann KK (2004) Reverse genetics of mononegavirales. In: Current Topics in Microbiology and Immunology (in press)

Conzelmann KK, Cox JH, Schneider LG, Thiel HJ (1990) Molecular cloning and complete nucleotide sequence of the attenuated rabies virus SAD B19. Virology 175:485–499

Coulon P, Ternaux JP, Flamand A, Tuffereau C (1998) An avirulent mutant of rabies virus is unable to infect motoneurons in vivo and in vitro. J Virol 72:273–278

Dalton KP, Rose JK (2001) Vesicular stomatitis virus glycoprotein containing the entire green fluorescent protein on its cytoplasmic domain is incorporated efficiently into virus particles. Virology 279:414–421

DeMattos CA, DeMattos CC, Rupprecht CE (2001) Rhabdoviridae: the viruses and their replication. In: Knipe DM, Howley PM (eds) Fields Virology. Vol. 1. Lippincott Williams & Wilkins, Philadelphia, pp 1245–1277

Dietzschold B, Wunner WH, Wiktor TJ, Lopes AD, Lafon M, Smith CL, Koprowski H (1983) Characterization of an antigenic determinant of the glycoprotein that correlates with pathogenicity of rabies virus. Proc Natl Acad Sci U S A 80:70–74

Drolet BS, Chiou PP, Heidel J, Leong JA (1995) Detection of truncated virus particles in a persistent RNA virus infection in vivo. J Virol 69:2140–2147

Etessami R, Conzelmann KK, Fadai-Ghotbi B, Natelson B, Tsiang H, Ceccaldi PE (2000) Spread and pathogenic characteristics of a G-deficient rabies virus recombinant: an in vitro and in vivo study. J Gen Virol 81:2147–2153

Ezelle HJ, Markovic D, Barber GN (2002) Generation of hepatitis C virus-like particles by use of a recombinant vesicular stomatitis virus vector. J Virol 76:12325–12334

Faber M, Pulmanausahakul R, Hodawadekar SS, Spitsin S, McGettigan JP, Schnell MJ, Dietzschold B (2002) Overexpression of the rabies virus glycoprotein results in enhancement of apoptosis and antiviral immune response. J Virol 76:3374–3381

Fernandez M, Porosnicu M, Markovic D, Barber GN (2002) Genetically engineered vesicular stomatitis virus in gene therapy: application for treatment of malignant disease. J Virol 76:895–904

Ferran MC, Lucas-Lenard JM (1997) The vesicular stomatitis virus matrix protein inhibits transcription from the human beta interferon promoter. J Virol 71:371–377

Finke S, Conzelmann KK (1997) Ambisense gene expression from recombinant rabies virus: random packaging of positive- and negative-strand ribonucleoprotein complexes into rabies virions. J Virol 71:7281–7288

Finke S, Conzelmann KK (1999) Virus promoters determine interference by defective RNAs: selective amplification of mini-RNA vectors and rescue from cDNA by a 3′ copy-back ambisense rabies virus. J Virol 73:3818–3825

Finke S, Conzelmann KK (2003) Dissociation of rabies virus matrix protein functions in regulation of viral RNA synthesis and virus assembly. J Virol 77:12074–12082

Finke S, Cox JH, Conzelmann KK (2000) Differential transcription attenuation of rabies virus genes by intergenic regions: generation of recombinant viruses overexpressing the polymerase gene. J Virol 74:7261–7269

Finke S, Mueller-Waldeck R, Conzelmann KK (2003) Rabies virus matrix protein regulates the balance of virus transcription and replication. J Gen Virol 84:1613–1621

Flanagan EB, Zamparo JM, Ball LA, Rodriguez LL, Wertz GW (2001) Rearrangement of the genes of vesicular stomatitis virus eliminates clinical disease in the natural host: new strategy for vaccine development. J Virol 75:6107–6114

Flanagan EB, Schoeb TR, Wertz GW (2003) Vesicular stomatitis viruses with rearranged genomes have altered invasiveness and neuropathogenesis in mice. J Virol 77:5740–5748

Foley HD, McGettigan JP, Siler CA, Dietzschold B, Schnell MJ (2000) A recombinant rabies virus expressing vesicular stomatitis virus glycoprotein fails to protect against rabies virus infection. Proc Natl Acad Sci U S A 97:14680–14685

Foley HD, Otero M, Orenstein JM, Pomerantz RJ, Schnell MJ (2002) Rhabdovirus-based vectors with human immunodeficiency virus type 1 (HIV-1) envelopes display HIV-1-like tropism and target human dendritic cells. J Virol 76:19–31

Fuerst TR, Niles EG, Studier FW, Moss B (1986) Eukaryotic transient-expression system based on recombinant vaccinia virus that synthesizes bacteriophage T7 RNA polymerase. Proc Natl Acad Sci U S A 83:8122–8126

Garoff H, Hewson R, Opstelten DJ (1998) Virus maturation by budding. Microbiol Mol Biol Rev 62:1171–1190

Grigera PR, Marzocca MP, Capozzo AV, Buonocore L, Donis RO, Rose JK (2000) Presence of bovine viral diarrhea virus (BVDV) E2 glycoprotein in VSV recombinant particles and induction of neutralizing BVDV antibodies in mice. Virus Res 69:3–15

Grote D, Russell SJ, Cornu TI, Cattaneo R, Vile R, Poland GA, Fielding AK (2001) Live attenuated measles virus induces regression of human lymphoma xenografts in immunodeficient mice. Blood 97:3746–3754

Haglund K, Forman J, Krausslich HG, Rose JK (2000) Expression of human immunodeficiency virus type 1 Gag protein precursor and envelope proteins from a vesicular stomatitis virus recombinant: high-level production of virus-like particles containing HIV envelope. Virology 268:112–121

Haglund K, Leiner I, Kerksiek K, Buonocore L, Pamer E, Rose JK (2002a) High-level primary CD8(+) T-cell response to human immunodeficiency virus type 1 gag and env generated by vaccination with recombinant vesicular stomatitis viruses. J Virol 76:2730–2738

Haglund K, Leiner I, Kerksiek K, Buonocore L, Pamer E, Rose JK (2002b) Robust recall and long-term memory T-cell responses induced by prime-boost regimens with heterologous live viral vectors expressing human immunodeficiency virus type 1 Gag and Env proteins. J Virol 76:7506–7517

Harty RN, Brown ME, Hayes FP, Wright NT, Schnell MJ (2001) Vaccinia virus-free recovery of vesicular stomatitis virus. J Mol Microbiol Biotechnol 3:513–517

Hinzman EE, Barr JN, Wertz GW (2002) Identification of an upstream sequence element required for vesicular stomatitis virus mRNA transcription. J Virol 76:7632–7641

Inoue K, Shoji Y, Kurane I, Iijima T, Sakai T, Morimoto K (2003) An improved method for recovering rabies virus from cloned cDNA. J Virol Methods 107:229–236

Ito N, Takayama M, Yamada K, Sugiyama M, Minamoto N (2001) Rescue of rabies virus from cloned cDNA and identification of the pathogenicity-related gene: glycoprotein gene is associated with virulence for adult mice. J Virol 75:9121–9128

Iverson LE, Rose JK (1981) Localized attenuation and discontinuous synthesis during vesicular stomatitis virus transcription. Cell 23:477–484

Jacob Y, Badrane H, Ceccaldi PE, Tordo N (2000) Cytoplasmic dynein LC8 interacts with lyssavirus phosphoprotein. J Virol 74:10217–10222

Jeetendra E, Robison CS, Albritton LM, Whitt MA (2002) The membrane-proximal domain of vesicular stomatitis virus g protein functions as a membrane fusion potentiator and can induce hemifusion. J Virol 76:12300–12311

Johnson JE, Schnell MJ, Buonocore L, Rose JK (1997) Specific targeting to CD4+ cells of recombinant vesicular stomatitis viruses encoding human immunodeficiency virus envelope proteins. J Virol 71:5060–5068

Johnson MC, Simon BE, Kim CH, Leong JA (2000) Production of recombinant snakehead rhabdovirus: the NV protein is not required for viral replication. J Virol 74:2343–2350

Kahn JS, Schnell MJ, Buonocore L, Rose JK (1999) Recombinant vesicular stomatitis virus expressing respiratory syncytial virus (RSV) glycoproteins: RSV fusion protein can mediate infection and cell fusion. Virology 254:81–91

Kahn JS, Roberts A, Weibel C, Buonocore L, Rose JK (2001) Replication-competent or attenuated, nonpropagating vesicular stomatitis viruses expressing respiratory syncytial virus (RSV) antigens protect mice against RSV challenge. J Virol 75:11079–11087

Khatchikian D, Orlich M, Rott R (1989) Increased viral pathogenicity after insertion of a 28S ribosomal RNA sequence into the haemagglutinin gene of an influenza virus. Nature 340:156–157

Kim CH, Dummer DM, Chiou PP, Leong JA (1999) Truncated particles produced in fish surviving infectious hematopoietic necrosis virus infection: mediators of persistence? J Virol 73:843–849

Klas SD, Robison CS, Whitt MA, Miller MA (2002) Adjuvanticity of an IL-12 fusion protein expressed by recombinant deltaG-vesicular stomatitis virus. Cell Immunol 218:59–73

Kretzschmar E, Buonocore L, Schnell MJ, Rose JK (1997) High-efficiency incorporation of functional influenza virus glycoproteins into recombinant vesicular stomatitis viruses. J Virol 71:5982–5989

Lafon M, Lafage M, Martinez-Arends A, Ramirez R, Vuillier F, Charron D, Lotteau V, Scott-Algara D (1992) Evidence for a viral superantigen in humans. Nature 358:507–510

Lafon M, Scott-Algara D, Marche PN, Cazenave PA, Jouvin-Marche E (1994) Neonatal deletion and selective expansion of mouse T cells by exposure to rabies virus nucleocapsid superantigen. J Exp Med 180:1207–1215

Lagging LM, Meyer K, Owens RJ, Ray R (1998) Functional role of hepatitis C virus chimeric glycoproteins in the infectivity of pseudotyped virus. J Virol 72:3539–3546

Lawson ND, Stillman EA, Whitt MA, Rose JK (1995) Recombinant vesicular stomatitis viruses from DNA. Proc Natl Acad Sci U S A 92:4477–4481

Le Mercier P, Garcin D, Hausmann S, Kolakofsky D (2002) Ambisense sendai viruses are inherently unstable but are useful to study viral RNA synthesis. J Virol 76:5492–5502

Letchworth GJ, Barrera JC, Fishel JR, Rodriguez L (1996) Vesicular stomatitis New Jersey virus RNA persists in cattle following convalescence. Virology 219:480–484

Li Y, Luo L, Schubert M, Wagner RR, Kang CY (1993) Viral liposomes released from insect cells infected with recombinant baculovirus expressing the matrix protein of vesicular stomatitis virus. J Virol 67:4415–4420

Lorence RM, Reichard KW, Katubig BB, Reyes HM, Phuangsab A, Mitchell BR, Cascino CJ, Walter RJ, Peeples ME (1994) Complete regression of human neuroblastoma xenografts in athymic mice after local Newcastle disease virus therapy. J Natl Cancer Inst 86:1228–1233

Lyles DS (2000) Cytopathogenesis and inhibition of host gene expression by RNA viruses. Microbiol Mol Biol Rev 64:709–724

Marcus PI, Rodriguez LL, Sekellick MJ (1998) Interferon induction as a quasispecies marker of vesicular stomatitis virus populations. J Virol 72:542–549
Martinez I, Rodriguez LL, Jimenez C, Pauszek SJ, Wertz GW (2003) Vesicular stomatitis virus glycoprotein is a determinant of pathogenesis in swine, a natural host. J Virol 77:8039–8047
Matsuura Y, Tani H, Suzuki K, Kimura-Someya T, Suzuki R, Aizaki H, Ishii K, Moriishi K, Robison CS, Whitt MA, Miyamura T (2001) Characterization of pseudotype VSV possessing HCV envelope proteins. Virology 286:263–275
McGettigan JP, Foley HD, Belyakov IM, Berzofsky JA, Pomerantz RJ, Schnell MJ (2001a) Rabies virus-based vectors expressing human immunodeficiency virus type 1 (HIV-1) envelope protein induce a strong, cross-reactive cytotoxic T-lymphocyte response against envelope proteins from different HIV-1 isolates. J Virol 75:4430–4434
McGettigan JP, Sarma S, Orenstein JM, Pomerantz RJ, Schnell MJ (2001b) Expression and immunogenicity of human immunodeficiency virus type 1 Gag expressed by a replication-competent rhabdovirus-based vaccine vector. J Virol 75:8724–8732
McGettigan JP, Naper K, Orenstein J, Koser M, McKenna PM, Schnell MJ (2003a) Functional human immunodeficiency virus type 1 (HIV-1) Gag-Pol or HIV-1 Gag-Pol and env expressed from a single rhabdovirus-based vaccine vector genome. J Virol 77:10889–10899
McGettigan JP, Pomerantz RJ, Siler CA, McKenna PM, Foley HD, Dietzschold B, Schnell MJ (2003b) Second-generation rabies virus-based vaccine vectors expressing human immunodeficiency virus type 1 gag have greatly reduced pathogenicity but are highly immunogenic. J Virol 77:237–244
Mebatsion T (2001) Extensive attenuation of rabies virus by simultaneously modifying the dynein light chain binding site in the P protein and replacing Arg333 in the G protein. J Virol 75:11496–11502
Mebatsion T, Conzelmann KK (1996) Specific infection of CD4+ target cells by recombinant rabies virus pseudotypes carrying the HIV-1 envelope spike protein. Proc Natl Acad Sci U S A 93:11366–11370
Mebatsion T, Schnell MJ, Conzelmann KK (1995) Mokola virus glycoprotein and chimeric proteins can replace rabies virus glycoprotein in the rescue of infectious defective rabies virus particles. J Virol 69:1444–1451
Mebatsion T, Konig M, Conzelmann KK (1996a) Budding of rabies virus particles in the absence of the spike glycoprotein. Cell 84:941–951
Mebatsion T, Schnell MJ, Cox JH, Finke S, Conzelmann KK (1996b) Highly stable expression of a foreign gene from rabies virus vectors. Proc Natl Acad Sci U S A 93:7310–7314
Mebatsion T, Finke S, Weiland F, Conzelmann KK (1997) A CXCR4/CD4 pseudotype rhabdovirus that selectively infects HIV-1 envelope protein-expressing cells. Cell 90:841–847
Mebatsion T, Weiland F, Conzelmann KK (1999) Matrix protein of rabies virus is responsible for the assembly and budding of bullet-shaped particles and interacts with the transmembrane spike glycoprotein G. J Virol 73:242–250
Miller MA, Lavine CL, Klas SD, Pfeffer LM, Whitt MA (2004) Recombinant replication-restricted VSV as an expression vector for murine cytokines. Protein Expr Purif 33:92–103
Miyoshi K, Harter DH, Hsu KC (1971) Neuropathological and immunofluorescence studies of experimental vesicular stomatitis virus encephalitis in mice. J Neuropathol Exp Neurol 30:266–277

Morimoto K, Hooper DC, Spitsin S, Koprowski H, Dietzschold B (1999) Pathogenicity of different rabies virus variants inversely correlates with apoptosis and rabies virus glycoprotein expression in infected primary neuron cultures. J Virol 73:510–518

Morimoto K, Foley HD, McGettigan JP, Schnell MJ, Dietzschold B (2000) Reinvestigation of the role of the rabies virus glycoprotein in viral pathogenesis using a reverse genetics approach. J Neurovirol 6:373–381

Morimoto K, McGettigan JP, Foley HD, Hooper DC, Dietzschold B, Schnell MJ (2001a) Genetic engineering of live rabies vaccines. Vaccine 19:3543–3551

Morimoto K, Schnell MJ, Pulmanausahakul R, McGettigan JP, Foley HD, Faber M, Hooper DC, Dietzschold B (2001b) High level expression of a human rabies virus-neutralizing monoclonal antibody by a rhabdovirus-based vector. J Immunol Methods 252:199–206

Neumann G, Whitt MA, Kawaoka Y (2002) A decade after the generation of a negative-sense RNA virus from cloned cDNA. What have we learned? J Gen Virol 83:2635–2662

Obuchi M, Fernandez M, Barber GN (2003) Development of recombinant vesicular stomatitis viruses that exploit defects in host defense to augment specific oncolytic activity. J Virol 77:8843–8856

Ogino M, Ebihara H, Lee BH, Araki K, Lundkvist A, Kawaoka Y, Yoshimatsu K, Arikawa J (2003) Use of vesicular stomatitis virus pseudotypes bearing hantaan or Seoul virus envelope proteins in a rapid and safe neutralization test. Clin Diagn Lab Immunol 10:154–160

Okuma K, Matsuura Y, Tatsuo H, Inagaki Y, Nakamura M, Yamamoto N, Yanagi Y (2001) Analysis of the molecules involved in human T-cell leukaemia virus type 1 entry by a vesicular stomatitis virus pseudotype bearing its envelope glycoproteins. J Gen Virol 82:821–830

Okuma K, Dalton KP, Buonocore L, Ramsburg E, Rose JK (2003) Development of a novel surrogate virus for human T-cell leukemia virus type 1: inhibition of infection by osteoprotegerin. J Virol 77:8562–8569

Orlich M, Gottwald H, Rott R (1994) Nonhomologous recombination between the hemagglutinin gene and the nucleoprotein gene of an influenza virus. Virology 204:462–465

Plyusnin A, Kukkonen SK, Plyusnina A, Vapalahti O, Vaheri A (2002) Transfection-mediated generation of functionally competent Tula hantavirus with recombinant S RNA segment. EMBO J 21:1497–1503

Poisson N, Real E, Gaudin Y, Vaney MC, King S, Jacob Y, Tordo N, Blondel D (2001) Molecular basis for the interaction between rabies virus phosphoprotein P and the dynein light chain LC8: dissociation of dynein-binding properties and transcriptional functionality of P. J Gen Virol 82:2691–2696

Porosnicu M, Mian A, Barber GN (2003) The oncolytic effect of recombinant vesicular stomatitis virus is enhanced by expression of the fusion cytosine deaminase/uracil phosphoribosyltransferase suicide gene. Cancer Res 63:8366–8376

Pringle CR (1997) The order Mononegavirales–current status. Arch Virol 142:2321–2326

Prosniak M, Faber M, Hanlon CA, Rupprecht CE, Hooper DC, Dietzschold B (2003) Development of a cocktail of recombinant-expressed human rabies virus-neutralizing monoclonal antibodies for postexposure prophylaxis of rabies. J Infect Dis 188:53–56

Pulmanausahakul R, Faber M, Morimoto K, Spitsin S, Weihe E, Hooper DC, Schnell MJ, Dietzschold B (2001) Overexpression of cytochrome C by a recombinant rabies virus attenuates pathogenicity and enhances antiviral immunity. J Virol 75:10800–10807

Quinones-Kochs MI, Schnell MJ, Buonocore L, Rose JK (2001) Mechanisms of loss of foreign gene expression in recombinant vesicular stomatitis viruses. Virology 287:427–435

Quiroz E, Moreno N, Peralta PH, Tesh RB (1988) A human case of encephalitis associated with vesicular stomatitis virus (Indiana serotype) infection. Am J Trop Med Hyg 39:312–314

Ramsburg E, Rose NF, Marx PA, Mefford M, Nixon DF, Moretto WJ, Montefiori D, Earl P, Moss B, Rose JK (2004) Highly effective control of an AIDS virus challenge in macaques by using vesicular stomatitis virus and modified vaccinia virus Ankara vaccine vectors in a single-boost protocol. J Virol 78:3930–3940

Raux H, Flamand A, Blondel D (2000) Interaction of the rabies virus P protein with the LC8 dynein light chain. J Virol 74:10212–10216

Reuter JD, Vivas-Gonzalez BE, Gomez D, Wilson JH, Brandsma JL, Greenstone HL, Rose JK, Roberts A (2002) Intranasal vaccination with a recombinant vesicular stomatitis virus expressing cottontail rabbit papillomavirus L1 protein provides complete protection against papillomavirus-induced disease. J Virol 76:8900–8909

Revilla T, Garci, Ramos G (2003) Fighting a virus with a virus: a dynamic model for HIV-1 therapy. Math Biosci 185:191–203

Roberts A, Kretzschmar E, Perkins AS, Forman J, Price R, Buonocore L, Kawaoka Y, Rose JK (1998) Vaccination with a recombinant vesicular stomatitis virus expressing an influenza virus hemagglutinin provides complete protection from influenza virus challenge. J Virol 72:4704–4711

Roberts A, Buonocore L, Price R, Forman J, Rose JK (1999) Attenuated vesicular stomatitis viruses as vaccine vectors. J Virol 73:3723–3732

Robison CS, Whitt MA (2000) The membrane-proximal stem region of vesicular stomatitis virus G protein confers efficient virus assembly. J Virol 74:2239–2246

Rose JK (1980) Complete intergenic and flanking gene sequences from the genome of vesicular stomatitis virus. Cell 19:415–421

Rose JK, Whitt MA (2001) Rhabdoviridae: the viruses and their replication. In: Knipe DM, Howley PM (eds) Fields virology. Vol. 1, Lippincott Williams & Wilkins, Philadelphia, pp 1221–1244

Rose NF, Roberts A, Buonocore L, Rose JK (2000) Glycoprotein exchange vectors based on vesicular stomatitis virus allow effective boosting and generation of neutralizing antibodies to a primary isolate of human immunodeficiency virus type 1. J Virol 74:10903–10910

Rose NF, Marx PA, Luckay A, Nixon DF, Moretto WJ, Donahoe SM, Montefiori D, Roberts A, Buonocore L, Rose JK (2001) An effective AIDS vaccine based on live attenuated vesicular stomatitis virus recombinants. Cell 106:539–549

Russell SJ (2002) RNA viruses as virotherapy agents. Cancer Gene Ther 9:961–966

Sabin A, Olitsky P (1938) Influence of host factors on neuroinvasiveness of vesicular stomatitis virus. I. Effect of age on the invasion of the brain by virus instilled in the nose. J Exp Med 66:15–34

Sakaguchi T, Uchiyama T, Fujii Y, Kiyotani K, Kato A, Nagai Y, Kawai A, Yoshida T (1999) Double-layered membrane vesicles released from mammalian cells infected with Sendai virus expressing the matrix protein of vesicular stomatitis virus. Virology 263:230–243

Schickli JH, Flandorfer A, Nakaya T, Martinez-Sobrido L, Garcia-Sastre A, Palese P (2001) Plasmid-only rescue of influenza A virus vaccine candidates. Philos Trans R Soc Lond B Biol Sci 356:1965–1973

Schirrmacher V, Ahlert T, Probstle T, Steiner HH, Herold-Mende C, Gerhards R, Hagmuller E, Steiner HH (1998) Immunization with virus-modified tumor cells. Semin Oncol 25:677–696

Schlereth B, Rose JK, Buonocore L, ter Muelen V, Niewiesk S (2000) Successful vaccine-induced seroconversion by single-dose immunization in the presence of measles virus-specific maternal antibodies. J Virol 74:4652–4657

Schlereth B, Buonocore L, Tietz A, Meulen Vt V, Rose JK, Niewiesk S (2003) Successful mucosal immunization of cotton rats in the presence of measles virus-specific antibodies depends on degree of attenuation of vaccine vector and virus dose. J Gen Virol 84:2145–2151

Schnell MJ (2001) Viral vectors as potential HIV-1 vaccines. FEMS Microbiol Lett 200:123–129

Schnell MJ, Mebatsion T, Conzelmann KK (1994) Infectious rabies viruses from cloned cDNA. EMBO J 13:4195–4203

Schnell MJ, Buonocore L, Kretzschmar E, Johnson E, Rose JK (1996a) Foreign glycoproteins expressed from recombinant vesicular stomatitis viruses are incorporated efficiently into virus particles. Proc Natl Acad Sci U S A 93:11359–11365

Schnell MJ, Buonocore L, Whitt MA, Rose JK (1996b) The minimal conserved transcription stop-start signal promotes stable expression of a foreign gene in vesicular stomatitis virus. J Virol 70:2318–2323

Schnell MJ, Johnson JE, Buonocore L, Rose JK (1997) Construction of a novel virus that targets HIV-1-infected cells and controls HIV-1 infection. Cell 90:849–857

Schnell MJ, Buonocore L, Boritz E, Ghosh HP, Chernish R, Rose JK (1998) Requirement for a non-specific glycoprotein cytoplasmic domain sequence to drive efficient budding of vesicular stomatitis virus. EMBO J 17:1289–1296

Schnell MJ, Foley HD, Siler CA, McGettigan JP, Dietzschold B, Pomerantz RJ (2000) Recombinant rabies virus as potential live-viral vaccines for HIV-1. Proc Natl Acad Sci U S A 97:3544–3549

Schnitzer TJ, Dickson C, Weiss RA (1979) Morphological and biochemical characterization of viral particles produced by the tsO45 mutant of vesicular stomatitis virus at restrictive temperature. J Virol 29:185–195

Siler CA, McGettigan JP, Dietzschold B, Herrine SK, Dubuisson J, Pomerantz RJ, Schnell MJ (2002) Live and killed rhabdovirus-based vectors as potential hepatitis C vaccines. Virology 292:24–34

Sinkovics JG (1991) Viral oncolysates as human tumor vaccines. Int Rev Immunol 7:259–287

Sinkovics JG, Horvath JC (2000) Newcastle disease virus (NDV): brief history of its oncolytic strains. J Clin Virol 16:1–15

Sodja I (1980) Nonlethal infection of laboratory mice induced with "mouse" rabies strains. Acta Virol 24:325–333

Spann KM, Collins PL, Teng MN (2003) Genetic recombination during coinfection of two mutants of human respiratory syncytial virus. J Virol 77:11201–11211

Stanziale SF, Fong Y (2003) Novel approaches to cancer therapy using oncolytic viruses. Curr Mol Med 3:61–71

Stillman EA, Whitt MA (1998) The length and sequence composition of vesicular stomatitis virus intergenic regions affect mRNA levels and the site of transcript initiation. J Virol 72:5565–5572

Stillman EA, Whitt MA (1999) Transcript initiation and 5′-end modifications are separable events during vesicular stomatitis virus transcription. J Virol 73:7199–7209

Stojdl DF, Lichty B, Knowles S, Marius R, Atkins H, Sonenberg N, Bell JC (2000) Exploiting tumor-specific defects in the interferon pathway with a previously unknown oncolytic virus. Nat Med 6:821–825

Takada A, Robison C, Goto H, Sanchez A, Murti KG, Whitt MA, Kawaoka Y (1997) A system for functional analysis of Ebola virus glycoprotein. Proc Natl Acad Sci U S A 94:14764–14769

Takada A, Feldmann H, Stroeher U, Bray M, Watanabe S, Ito H, McGregor M, Kawaoka Y (2003) Identification of protective epitopes on Ebola virus glycoprotein at the single amino acid level by using recombinant vesicular stomatitis viruses. J Virol 77:1069–1074

Tatsuo H, Okuma K, Tanaka K, Ono N, Minagawa H, Takade A, Matsuura Y, Yanagi Y (2000) Virus entry is a major determinant of cell tropism of Edmonston and wild-type strains of measles virus as revealed by vesicular stomatitis virus pseudotypes bearing their envelope proteins. J Virol 74:4139–4145

Tesh RB, Peralta PH, Johnson KM (1969) Ecologic studies of vesicular stomatitis virus. I. Prevalence of infection among animals and humans living in an area of endemic VSV activity. Am J Epidemiol 90:255–261

Tordo N, Poch O, Ermine A, Keith G, Rougeon F (1986) Walking along the rabies genome: is the large G-L intergenic region a remnant gene? Proc Natl Acad Sci U S A 83:3914–3918

Wallack MK, Sivanandham M, Balch CM, Urist MM, Bland KI, Murray D, Robinson WA, Flaherty L, Richards JM, Bartolucci AA, Rosen L (1998) Surgical adjuvant active specific immunotherapy for patients with stage III melanoma: the final analysis of data from a phase III, randomized, double-blind, multicenter vaccinia melanoma oncolysate trial. J Am Coll Surg 187:69–77

Wertz GW, Perepelitsa VP, Ball LA (1998) Gene rearrangement attenuates expression and lethality of a nonsegmented negative strand RNA virus. Proc Natl Acad Sci U S A 95:3501–3506

Wertz GW, Moudy R, Ball LA (2002) Adding genes to the RNA genome of vesicular stomatitis virus: positional effects on stability of expression. J Virol 76:7642–7650

Whelan SP, Ball LA, Barr JN, Wertz GT (1995) Efficient recovery of infectious vesicular stomatitis virus entirely from cDNA clones. Proc Natl Acad Sci U S A 92:8388–8392

Whelan SP, Wertz GW (1999) The 5′ terminal trailer region of vesicular stomatitis virus contains a position-dependent cis-acting signal for assembly of RNA into infectious particles. J Virol 73:307–315

Whitt MA, Hnatyszyn HJ, Spruill G, Robison CS, Barnes JD, Jayakar HR, Bauler M, Watanabe M, Sherman P, Allay JA, Barber GN, Steiner MS (2003) Reduction of HIV load and restoration of CD4+ T cells by a novel anti-HIV recombinant vesicular stomatitis virus cytolytic agent (GTx-v311) XII International Conference on Negative Strand Viruses, June 14th–19th 2003, Pisa, Italy

Zinkernagel RM (1997) Felix Hoppe-Seyler Lecture 1997. Protective antibody responses against viruses. Biol Chem 378:725–729

Subject Index

Current Topics in Microbiology and Immunology

Volumes published since 1989 (and still available)

Vol. 247: **Wagner, Hermann (Ed.):** Immunobiology of Bacterial CpGDNA. 2000. 34 figs. IX, 246 pp. ISBN 3-540-66400-9

Vol. 248: **du Pasquier, Louis; Litman, Gary W. (Eds.):** Origin and Evolution of the Vertebrate Immune System. 2000. 81 figs. IX, 324 pp. ISBN 3-540-66414-9

Vol. 249: **Jones, Peter A.; Vogt, Peter K. (Eds.):** DNA Methylation and Cancer. 2000. 16 figs. IX, 169 pp. ISBN 3-540-66608-7

Vol. 250: **Aktories, Klaus; Wilkins, Tracy, D. (Eds.):** Clostridium difficile. 2000. 20 figs. IX, 143 pp. ISBN 3-540-67291-5

Vol. 251: **Melchers, Fritz (Ed.):** Lymphoid Organogenesis. 2000. 62 figs. XII, 215 pp. ISBN 3-540-67569-8

Vol. 252: **Potter, Michael; Melchers, Fritz (Eds.):** B1 Lymphocytes in B Cell Neoplasia. 2000. XIII, 326 pp. ISBN 3-540-67567-1

Vol. 253: **Gosztonyi, Georg (Ed.):** The Mechanisms of Neuronal Damage in Virus Infections of the Nervous System. 2001. approx. XVI, 270 pp. ISBN 3-540-67617-1

Vol. 254: **Privalsky, Martin L. (Ed.):** Transcriptional Corepressors. 2001. 25 figs. XIV, 190 pp. ISBN 3-540-67569-8

Vol. 255: **Hirai, Kanji (Ed.):** Marek's Disease. 2001. 22 figs. XII, 294 pp. ISBN 3-540-67798-4

Vol. 256: **Schmaljohn, Connie S.; Nichol, Stuart T. (Eds.):** Hantaviruses. 2001. 24 figs. XI, 196 pp. ISBN 3-540-41045-7

Vol. 257: **van der Goot, Gisou (Ed.):** Pore-Forming Toxins, 2001. 19 figs. IX, 166 pp. ISBN 3-540-41386-3

Vol. 258: **Takada, Kenzo (Ed.):** Epstein-Barr Virus and Human Cancer. 2001. 38 figs. IX, 233 pp. ISBN 3-540-41506-8

Vol. 259: **Hauber, Joachim, Vogt, Peter K. (Eds.):** Nuclear Export of Viral RNAs. 2001. 19 figs. IX, 131 pp. ISBN 3-540-41278-6

Vol. 260: **Burton, Didier R. (Ed.):** Antibodies in Viral Infection. 2001. 51 figs. IX, 309 pp. ISBN 3-540-41611-0

Vol. 261: **Trono, Didier (Ed.):** Lentiviral Vectors. 2002. 32 figs. X, 258 pp. ISBN 3-540-42190-4

Vol. 262: **Oldstone, Michael B.A. (Ed.):** Arenaviruses I. 2002. 30 figs. XVIII, 197 pp. ISBN 3-540-42244-7

Vol. 263: **Oldstone, Michael B. A. (Ed.):** Arenaviruses II. 2002. 49 figs. XVIII, 268 pp. ISBN 3-540-42705-8

Vol. 264/I: **Hacker, Jörg; Kaper, James B. (Eds.):** Pathogenicity Islands and the Evolution of Microbes. 2002. 34 figs. XVIII, 232 pp. ISBN 3-540-42681-7

Vol. 264/II: **Hacker, Jörg; Kaper, James B. (Eds.):** Pathogenicity Islands and the Evolution of Microbes. 2002. 24 figs. XVIII, 228 pp. ISBN 3-540-42682-5

Vol. 265: **Dietzschold, Bernhard; Richt, Jürgen A. (Eds.):** Protective and Pathological Immune Responses in the CNS. 2002. 21 figs. X, 278 pp. ISBN 3-540-42668-X

Vol. 266: **Cooper, Koproski (Eds.):** The Interface Between Innate and Acquired Immunity, 2002. 15 figs. XIV, 116 pp. ISBN 3-540-42894-X

Vol. 267: **Mackenzie, John S.; Barrett, Alan D. T.; Deubel, Vincent (Eds.):** Japanese Encephalitis and West Nile Viruses. 2002. 66 figs. X, 418 pp. ISBN 3-540-42783-X

Vol. 268: **Zwickl, Peter; Baumeister, Wolfgang (Eds.):** The Proteasome-Ubiquitin Protein Degradation Pathway. 2002. 17 figs. X, 213 pp. ISBN 3-540-43096-2

Vol. 269: **Koszinowski, Ulrich H.; Hengel, Hartmut (Eds.):** Viral Proteins Counteracting Host Defenses. 2002. 47 figs. XII, 325 pp. ISBN 3-540-43261-2

Vol. 270: **Beutler, Bruce; Wagner, Hermann (Eds.):** Toll-Like Receptor Family Members and Their Ligands. 2002. 31 figs. X, 192 pp. ISBN 3-540-43560-3

Vol. 271: **Koehler, Theresa M. (Ed.):** Anthrax. 2002. 14 figs. X, 169 pp. ISBN 3-540-43497-6

Vol. 272: **Doerfler, Walter; Böhm, Petra (Eds.):** Adenoviruses: Model and Vectors in Virus-Host Interactions. Virion and Structure, Viral Replication, Host Cell Interactions. 2003. 63 figs., approx. 280 pp. ISBN 3-540-00154-9

Vol. 273: **Doerfler, Walter; Böhm, Petra (Eds.):** Adenoviruses: Model and Vectors in VirusHost Interactions. Immune System, Oncogenesis, Gene Therapy. 2004. 35 figs., approx. 280 pp. ISBN 3-540-06851-1

Vol. 274: **Workman, Jerry L. (Ed.):** Protein Complexes that Modify Chromatin. 2003. 38 figs., XII, 296 pp. ISBN 3-540-44208-1

Vol. 275: **Fan, Hung (Ed.):** Jaagsiekte Sheep Retrovirus and Lung Cancer. 2003. 63 figs., XII, 252 pp. ISBN 3-540-44096-3

Vol. 276: **Steinkasserer, Alexander (Ed.):** Dendritic Cells and Virus Infection. 2003. 24 figs., X, 296 pp. ISBN 3-540-44290-1

Vol. 277: **Rethwilm, Axel (Ed.):** Foamy Viruses. 2003. 40 figs., X, 214 pp. ISBN 3-540-44388-6

Vol. 278: **Salomon, Daniel R.; Wilson, Carolyn (Eds.):** Xenotransplantation. 2003. 22 figs., IX, 254 pp. ISBN 3-540-00210-3

Vol. 279: **Thomas, George; Sabatini, David; Hall, Michael N. (Eds.):** TOR. 2004. 49 figs., X, 364 pp. ISBN 3-540-00534-X

Vol. 280: **Heber-Katz, Ellen (Ed.):** Regeneration: Stem Cells and Beyond. 2004. 42 figs., XII, 194 pp. ISBN 3-540-02238-4

Vol. 281: **Young, John A. T. (Ed.):** Cellular Factors Involved in Early Steps of Retroviral Replication. 2003. 21 figs., IX, 240 pp. ISBN 3-540-00844-6

Vol. 282: **Stenmark, Harald (Ed.):** Phosphoinositides in Subcellular Targeting and Enzyme Activation. 2003. 20 figs., X, 210 pp. ISBN 3-540-00950-7

Vol. 283: **Kawaoka, Yoshihiro (Ed.):** Biology of Negative Strand RNA Viruses: The Power of Reverse Genetics. 2004. 24 figs., IX, 350 pp. ISBN 3-540-40661-1

Vol. 284: **Harris, David (Ed.):** Mad Cow Disease and Related Spongiform Encephalopathies. 2004. 34 figs., IX, 219 pp. ISBN 3-540-20107-6

Vol. 285: **Marsh, Mark (Ed.):** Membrane Trafficking in Viral Replication. 2004. 19 figs., IX, 259 pp. ISBN 3-540-21430-5

Vol. 286: **Madshus, Inger H. (Ed.):** Signalling from Internalized Growth Factor Receptors. 2004. 19 figs., IX, 187 pp. ISBN 3-540-21038-5

Vol. 287: **Enjuanes, Luis (Ed.):** Coronavirus Replication and Reverse Genetics. 2005. 49 figs., XI, 257 pp. ISBN 3-540-21494-1

Vol. 288: **Mahy, Brain W. J. (Ed.):** Foot-and-Mouth-Disease Virus. 2005. 16 figs., IX, 178 pp. ISBN 3-540-22419-X

Vol. 289: **Griffin, Diane E. (Ed.):** Role of Apoptosis in Infection. 2005. 40 figs., IX, 294 pp. ISBN 3-540-23006-8

Vol. 290: **Singh, Harinder; Grosschedl, Rudolf (Eds.):** Molecular Analysis of B Lymphocyte Development and Activation. 2005. 28 figs., XI, 255 pp. ISBN 3-540-23090-4

Vol. 291: **Boquet, Patrice; Lemichez, Emmanuel (Eds.):** Bacterial Virulence Factors and Rho GTPases. 2005. 28 figs., IX, 196 pp. ISBN 3-540-23865-4

Printing: Mercedes-Druck, Berlin
Binding: Stein+Lehmann, Berlin